Differential Forms

Henri Cartan

DOVER PUBLICATIONS, INC.
Mineola, New York

Bibliographical Note

This Dover edition, first published in 2006, is an unabridged republication of the work originally published in 1971 by Kershaw Publishing Company Limited, London. It is published by special arrangement with Editions Hermann, 6 rue de la Sorbonne, 75005 Paris, France.

This is the second part of a course given in 1965/66 at the Faculty of Sciences, Paris, under the title *Calcul différentiel* in the series Méthodes.

Translated from the original French text *Formes différentielles,* first published by Hermann in 1967

The exercises are by Mme. C. Buttin, F. Ridear and J. L. Verley.

International Standard Book Number: 0-486-45010-4

Manufactured in the United States of America
Dover Publications, Inc., 31 East 2nd Street, Mineola, N.Y. 11501

Contents

Chapter 1. Differential forms

Chapter 2. Elements of the calculus of variations

Chapter 3. Applications of the moving frame method to the theory of curves and surfaces

Chapter 1

Differential forms

1. Multilinear alternating mappings

1.1 *Definition of an alternating multilinear mapping*

Much of the following will be of interest purely from the point of view of that division of algebra which is concerned with the study of vector spaces over an arbitrary commutative field K (no restriction being placed on the characteristic of the field). However, we shall confine ourselves in exposition to normed vector spaces over the fields **R** or **C**: to illustrate, let us take **R**.

Suppose then that E, F are two normed vector spaces. We have already considered (Chap. 1, § 1.8, *Differential Calculus*) the normed vector space $\mathscr{L}_p(E; F)$ of the p-linear continuous mappings $E^p \to F$; for $p = 1$ we write for simplicity $\mathscr{L}(E; F)$; for $p = 0$ we define $\mathscr{L}_0(E; F) = F$. We also considered the vector subspace of $\mathscr{L}_p(E; F)$ formed by the p-linear *symmetric* mappings. We shall now introduce a second subspace $\mathscr{A}_p(E; F)$ of $\mathscr{L}_p(E; F)$.

DEFINITION. A mapping $f \in \mathscr{L}_p(E; F)$ is called *alternating* if its value $f(x_1, \ldots, x_p)$ is null whenever $x_i = x_{i+1}$ for at least one i $(1 \leqslant i < p)$. (We agree that for $p = 1$ every linear function $E \to F$ is an alternation.)

It is evident that the p-linear alternating mappings form a vector subspace of $\mathscr{L}_p(E; F)$; we denote it by $\mathscr{A}_p(E; F)$. Then $\mathscr{A}_1(E; F) = \mathscr{L}(E; F)$; by definition we put $\mathscr{A}_0(E; F) = F$.

The vector subspace $\mathscr{A}_p(E; F)$ is *closed* in $\mathscr{L}_p(E; F)$; indeed, suppose $f \in \mathscr{L}_p(E; F)$ is the limit of a sequence $f_n \in \mathscr{A}_p(E; F)$: then $\lim_{n\to\infty} \|f - f_n\| = 0$; *a fortiori*, for $x_1, \ldots, x_p \in E$, the limit of $f_n(x_1, \ldots, x_p)$ is $f(x_1, \ldots, x_p)$. Thus if $x_i = x_{i+1}$, it follows that

$$f(x_1, \ldots, x_p) = \lim_n f_n(x_1, \ldots, x_p) = 0. \qquad \text{Q.E.D.}$$

Before stating further properties of multilinear alternating mappings we require a few preliminaries concerning permutation groups.

1.2 *Permutation groups*

Let Σ_p denote the group of all permutations of the set $\{1, 2, \ldots, p\}$ of the first p integers

>0. We know that it contains $p!$ elements. A *transposition* is a permutation $\sigma \in \Sigma_p$ such that there exist integers i, j $(i \neq j, 1 \leqslant i \leqslant p, 1 \leqslant j \leqslant p)$ for which

$$\begin{cases} \sigma(i) = j, & \sigma(j) = i \\ \sigma(k) = k & \text{for} \quad k \neq i, j. \end{cases}$$

(Note that σ interchanges i and j.) It is evident that σ^2 is the identity element. It can be shown that any permutation $\sigma \in \Sigma_p$ may be written as a product of transpositions, each of which interchanges two *consecutive* integers; further (see any elementary text on the theory of determinants), irrespective of the manner in which σ is written as a product of transpositions, the parity of the number of these transpositions is always the same.

The *signature* of a permutation σ is denoted by $\epsilon(\sigma)$, and is equal to $+1$ if σ is the product of an even number of transpositions, and -1 if σ is the product of an odd number of transpositions. Thus:

The mapping $\sigma \to \epsilon(\sigma)$ of the group Σ_p into the multiplicative group of the two elements $+1$ and -1 is a homomorphism; and if σ is a transposition, then $\epsilon(\sigma) = -1$.

Suppose that E and F are two sets, and let f denote the mapping $\underbrace{\mathrm{E} \times \cdots \times \mathrm{E}}_{p \text{ times}} \to \mathrm{F}$.

For $\sigma \in \Sigma_p$ we denote by σf the mapping of $\mathrm{E} \times \cdots \times \mathrm{E} \to \mathrm{F}$ defined by

$$(1.2.1) \qquad (\sigma f)(x_1, \ldots, x_p) = f(x_{\sigma(1)}, \ldots, x_{\sigma(p)})$$

(i.e. σf is obtained from f by a permutation of the variables). Clearly, if σ is the identity, then $\sigma f = f$. Further we have

$$(1.2.2) \qquad (\tau\sigma)f = \tau(\sigma f) \quad \text{for} \quad \sigma \in \Sigma_p, \tau \in \Sigma_p.$$

Indeed, putting $\sigma f = g$, we have

$$(\tau g)(x_1, \ldots, x_p) = g(x_{\tau(1)}, \ldots, x_{\tau(p)}).$$

Put $y_i = x_{\tau(i)}$; then by (1.2.1),

$$g(y_1, \ldots, y_p) = f(y_{\sigma(1)}, \ldots, y_{\sigma(p)}),$$

and since $y_{\sigma(i)} = x_{\tau\sigma(i)}$ it follows that $\tau(\sigma f)$ is the mapping

$$(x_1, \ldots, x_p) \mapsto f(x_{\tau\sigma(1)}, \ldots, x_{\tau\sigma(p)}).$$

This proves (1.2.2).

This relation shows that *the group Σ_p operates on the left in the set of mappings* $\mathrm{E}^p \to \mathrm{F}$.

1.3 *Properties of multilinear alternating mappings*

In the following, E and F will denote two normed vector spaces.

PROPOSITION 1.3.1. *Let $f \in \mathscr{L}_p(\mathrm{E}; \mathrm{F})$ be a multilinear alternating mapping; then:* (i) *each time $x_i = x_j$ for a pair (i, j) of distinct suffixes it is true that $f(x_1, \ldots, x_p) = 0$;* (ii) *for each permutation σ of $\{1, \ldots, p\}$ we have*

$$(1.3.1) \qquad f(x_{\sigma(1)}, \ldots, x_{\sigma(p)}) = \epsilon(\sigma) f(x_1, \ldots, x_p),$$

where $\epsilon(\sigma) = \pm 1$ is the signature of the permutation.

Before giving the proof let us note the following points: (i) generalizes the property that has served to *define* alternating mappings; it expresses the fact that $f(x_1, \ldots, x_p) = 0$ *as soon as two of the variables take the same value.*

PROOF. We shall prove (ii) first. To begin with, let us show that the relation (1.3.1) is true whenever σ is a *transposition* which interchanges two *consecutive* suffixes $i, i + 1$; i.e., that

$$g(x_{i+1}, x_i) = -g(x_i, x_{i+1}), \tag{1.3.2}$$

where for simplicity we have set

$$g(x_i, x_{i+1}) = g(x_1, \ldots, x_i, x_{i+1}, \ldots, x_p).$$

The function g is bilinear and alternating, therefore,

$$g(x_i + x_{i+1}, x_i + x_{i+1}) = g(x_i, x_i) + g(x_{i+1}, x_{i+1}) + g(x_i, x_{i+1}) + g(x_{i+1}, x_i)$$

and since the first three terms of this equality are null the result (1.3.2) follows.

Consider now the general case of (1.3.1); in the notation of § 1.2 it may be written

$$\sigma f = \epsilon(\sigma) f; \tag{1.3.3}$$

since $(\sigma_1\sigma_2)f = \sigma_1(\sigma_2 f)$, and $\epsilon(\sigma_1\sigma_2) = \epsilon(\sigma_1)\epsilon(\sigma_2)$, we see that if (1.3.3) is true for $\sigma = \sigma_1$ and $\sigma = \sigma_2$, it is true for $\sigma = \sigma_1\sigma_2$. Now it has been shown to be true when σ is a transposition which interchanges two consecutive suffixes, thus it is true for all finite products of such transpositions; and since every permutation may be written as such a product, it is therefore true for arbitrary $\sigma \in \Sigma_p$. This proves (ii).

It remains to prove (i). Suppose $x_i = x_j (i \neq j)$; there exists a permutation σ such that $\sigma(1) = i$, $\sigma(2) = j$; since f is alternating it is true that

$$f(x_{\sigma(1)}, \ldots, x_{\sigma(p)}) = 0,$$

and since this is equal to $\pm f(x_1, \ldots, x_p)$, we must have $f(x_1, \ldots, x_p) = 0$.

Q.E.D.

Remark. The above may be applied to vector spaces E and F defined over any field. But when the characteristic of K $\neq 2$ (which is certainly the case for **R** or **C**), *it follows from* (ii) *that a multilinear mapping is an alternating mapping.* In fact, equation (1.3.1), for the case where σ is a transposition interchanging i and j, shows that if $x_i = x_{i+1}$, then

$$f(x_1, \ldots, x_p) = -f(x_1, \ldots, x_p),$$

so that

$$2f(x_1, \ldots, x_p) = 0;$$

since we can divide by 2 we indeed obtain $f(x_1, \ldots, x_p) = 0$.

A multilinear mapping f possessing property (ii) is called *antisymmetric*; this is written briefly as $\sigma f = \epsilon(\sigma) f$.

Thus, in the cases of interest (K $=$ **R** or **C**), the space $\mathscr{A}_p(\mathrm{E}; \mathrm{F})$ of multilinear alternating mappings is none other than the space of multilinear antisymmetric mappings.

1.4 *Multiplication of multilinear alternating mappings*

Let $f \in \mathscr{A}_p(\mathrm{E}; \mathrm{F})$ and $g \in \mathscr{A}_q(\mathrm{E}; \mathrm{G})$. To define a multiplication between f and g we must first introduce a bilinear continuous mapping

$$\Phi: \mathrm{F} \times \mathrm{G} \to \mathrm{H}$$

(with values in a normed vector space H). This permits f and g to be associated with a mapping $h: \mathrm{E}^{p+q} \to \mathrm{H}$, viz.,

$$h(x_1, \ldots, x_{p+q}) = \Phi(f(x_1, \ldots, x_p), g(x_{p+1}, \ldots, x_{p+q})). \tag{1.4.1}$$

Evidently h is multilinear and continuous. But it is not in general alternating; it is only a vector space of $(p + q)$-linear mappings which are alternating when considered as a function of the first p variables $x_1, \ldots, x_p$, and alternating when considered as a function of the last q variables $x_{p+1}, \ldots, x_{p+q}$. We shall denote this space by $\mathscr{A}_{p,q}(\mathrm{E}; \mathrm{H})$.

We shall indicate a canonical procedure for associating with every $h \in \mathscr{A}_{p,q}(\mathrm{E}; \mathrm{H})$ an $\tilde{h} \in \mathscr{A}_{p+q}(\mathrm{E}; \mathrm{H})$. More precisely, we shall define a bilinear continuous mapping

$$\varphi_{p,q}: \mathscr{A}_{p,q}(\mathrm{E}; \mathrm{H}) \to \mathscr{A}_{p+q}(\mathrm{E}; \mathrm{H}).$$

By definition, $\varphi_{p+q}(h)$ is the multilinear mapping

$$\tilde{h} = \sum_{\sigma} \epsilon(\sigma)(\sigma h), \tag{1.4.2}$$

the summation being over all permutations σ of $\{1, \ldots, p + q\}$ such that

$$\sigma(1) < \cdots < \sigma(p) \quad \text{and} \quad \sigma(p+1) < \cdots < \sigma(p + q). \tag{1.4.3}$$

An intuitive idea of such a permutation is provided by the following example. Divide a pack of cards into two piles, the first having p cards and the second q cards (those of the first may be numbered 1 to p, and those of the second $p + 1$ to $p + q$). Now "shuffle" the pack by placing the second pile on top of the first; the cards are then arranged so that each of the two piles retains its initial ordering. In other words, the "shuffle" defines a permutation σ on the pack of cards which satisfies (1.4.3). And conversely. We observe that there are

$$\frac{(p + q)!}{p!q!}$$

such permutations.

We must now show that the mapping $\tilde{h}$ defined by (1.4.2) is indeed an *alternation*. Given a sequence of vectors $x_1, \ldots, x_{p+q} \in \mathrm{E}$ in which two consecutive x's are equal: $x_i = x_{i+1}$; it is required to show that $h(x_1, \ldots, x_{p+q})$ is null, i.e., that

$$\sum_{\sigma} \epsilon(\sigma) h(x_{\sigma(1)}, \ldots, x_{\sigma(p+q)}) = 0. \tag{1.4.4}$$

For clarity let us divide the permutations σ satisfying (1.4.3) into two categories:

1. Those σ for which $\sigma^{-1}(i)$ and $\sigma^{-1}(i+1)$ are both integers $\leqslant p$ or both $\geqslant p + 1$. In the first case x_i, x_{i+1} occur amongst the first p places in $h(x_{\sigma(1)}, \ldots, x_{\sigma(p+q)})$, therefore the corresponding term in (1.4.4) is null (h being an *alternating* mapping of the first p variables.) In the second case the corresponding term is also null for an analogous reason.

2. The other category is itself divided into two sub-categories: those σ for which $\sigma^{-1}(i) \leqslant p$ and $\sigma^{-1}(i+1) \geqslant p+1$, and those σ for which $\sigma^{-1}(i) \geqslant p+1$ and $\sigma^{-1}(i+1) \leqslant p$. Let τ be the *transposition* which interchanges i and $i+1$. If σ is in the first sub-category, $\tau\sigma$ is in the second, and vice versa. We may therefore group in pairs the remaining terms on the left-hand side of (1.4.4); for each σ such that $\sigma^{-1}(i) \leqslant p$ and $\sigma^{-1}(i+1) \geqslant p+1$ we take

$$\epsilon(\sigma)h(x_{\sigma(1)}, \ldots, x_{\sigma(p+q)}) - \epsilon(\sigma)h(x_{\tau\sigma(1)}, \ldots, x_{\tau\sigma(p+q)}). \tag{1.4.5}$$

Let us show that this is null, which will complete the proof of (1.4.4): the sequence $\tau\sigma(1), \ldots, \tau\sigma(p+q)$ is in fact deduced from $\sigma(1), \ldots, \sigma(p+q)$ by interchanging i and $i+1$, and since $x_i = x_{i+1}$, the value of (1.4.5) is indeed 0.

Having thus defined a canonical linear mapping

$$\varphi_{p,q} = \mathscr{A}_{p,q}(E; H) \to \mathscr{A}_{p+q}(E; H)$$

by formula (1.4.2) we may now give the

DEFINITION. *The exterior product of* $f \in \mathscr{A}_p(E; F)$ *and of* $g \in \mathscr{A}_q(E; G)$ *relative to* $\Phi: F \times G \to H$, *denoted by*

$$f \underset{\Phi}{\wedge} g,$$

is the element $\varphi_{p,q}(h) \in \mathscr{A}_{p+q}(E; H)$, where h is the element of $\mathscr{A}_{p,q}(E; H)$ defined by (1.4.1). As a formula:

$$(f \underset{\Phi}{\wedge} g)(x_1, \ldots, x_{p+q}) = \sum_{\sigma} \epsilon(\sigma)\Phi(f(x_{\sigma(1)}, \ldots, x_{\sigma(p)}), g(x_{\sigma(p+1)}, \ldots, x_{\sigma(p+q)})) \tag{1.4.6}$$

where the summation is over all permutations σ of $\{1, \ldots, p+q\}$ satisfying (1.4.3) (think of the pack of cards!).

Examples. Take the case $p = q = 1$; $f \in \mathscr{L}(E; F)$ and $g \in \mathscr{L}(E; G)$ are *linear* mappings. Then $f \underset{\Phi}{\wedge} g \in \mathscr{A}_2(E; H)$ is the bilinear mapping

$$(x_1, x_2) \to \Phi(f(x_1), g(x_2)) - \Phi(f(x_2), g(x_1)). \tag{1.4.7}$$

It is immediately clear in this simple case that it is an *alternation*, the right-hand side being zero when $x_1 = x_2$.

More generally, if $p = 1$ and q is arbitrary, then

$$(f \underset{\Phi}{\wedge} g)(x_0, x_1, \ldots, x_q) = \sum_{i=0}^{q} (-1)^i \Phi(f(x_i), g(x_0, \ldots, x_{i-1}, x_{i+1}, \ldots, x_q)) \tag{1.4.8}$$

We shall use the notation

$$g(x_0, \ldots, \hat{x}_i, \ldots, x_q)$$

to indicate that x_i has been removed from the sequence $x_1, \ldots, x_i, \ldots, x_q$.

A simple case is that in which $p = 0$, for which

$$(f \underset{\Phi}{\wedge} g)(x_1, \ldots, x_q) = \Phi(f, g(x_1, \ldots, x_q))$$

for $f \in F$ is a constant.

We shall often employ the multiplication $f \underset{\Phi}{\wedge} g$ when $G = \mathbf{R}$, $H = F$, the mapping

$\Phi: F \times \mathbf{R} \to F$ being simply the multiplication of a vector of F by a scalar. In this case, the Φ will be omitted in the notation $f \wedge g$. More particularly, if $F = \mathbf{R}$, the product $f \wedge g$ of two multilinear alternating forms will be a multilinear alternating form.

Again, more particularly, if $f \in \mathscr{L}(E; \mathbf{R})$ and $g \in \mathscr{L}(E; \mathbf{R})$ are two linear forms, their product $f \wedge g$ is an element of $\mathscr{A}_2(E; \mathbf{R})$ which, by (1.4.7), is given by

(1.4.9) $$(f \wedge g)(x_1, x_2) = f(x_1)g(x_2) - f(x_2)g(x_1).$$

Consider the mapping

$$(f, g) \mapsto f \wedge g$$

of $\mathscr{L}(E; \mathbf{R}) \times \mathscr{L}(E; \mathbf{R})$ into $\mathscr{A}_2(E; \mathbf{R})$. It is *bilinear alternating*; the bilinearity is obvious; it is alternating because, if $f = g$, the relation (1.4.9) shows that $f \wedge g = 0$, since scalar multiplication is commutative:

$$f(x_1)f(x_2) - f(x_2)f(x_1) = 0.$$

1.5 *Properties of exterior multiplication*

The mapping $(f, g) \mapsto f \underset{\Phi}{\wedge} g$ is *bilinear*: if g is fixed, $f \underset{\Phi}{\wedge} g$ depends linearly on f, if f is fixed, $f \underset{\Phi}{\wedge} g$ depends linearly on g. These assertions follow easily from (1.4.6).

PROPOSITION 1.5.1. *Let* $f \in \mathscr{A}_p(E; \mathbf{R})$ *and* $g \in \mathscr{A}_q(E; \mathbf{R})$ *be two multilinear alternating forms* (with scalar values). *Then*

(1.5.1) $$g \wedge f = (-1)^{pq} f \wedge g$$

(we say that exterior multiplication of alternating forms is *anticommutative*).

PROOF. We have

(1.5.2) $$(g \wedge f)(x_1, \ldots, x_{p+q}) = \sum_\tau \epsilon(\tau) g(x_{\tau(1)}, \ldots, x_{\tau(q)}) f(x_{\tau(q+1)}, \ldots, x_{\tau(p+q)}),$$

where the summation is over permutations τ of $\{1, \ldots, p+q\}$ satisfying

(1.5.3) $$\tau(1) < \cdots < \tau(q), \qquad \tau(q+1) < \cdots < \tau(q+p).$$

Introduce the permutation α which transforms the sequence $\{1, \ldots, p+q\}$ into

$$\{q+1, \ldots, q+p, 1, \ldots, p\}.$$

Then for $1 \leqslant i \leqslant q$, $\tau(i) = \tau\alpha(p+i)$, and for $q+1 \leqslant j \leqslant q+p$, $\tau(j) = \tau\alpha(j-q)$.

Putting $\tau\alpha = \sigma$, (1.5.3) implies that σ satisfies (1.4.3). Conversely, if σ satisfies (1.4.3) then $\tau = \sigma\alpha^{-1}$ satisfies (1.5.3). Moreover, $\epsilon(\tau) = \epsilon(\sigma)\epsilon(\alpha)$, and $\epsilon(\alpha) = (-1)^{pq}$, since to accomplish the transformation α one must interchange successively $1, \ldots, q$ with $q+1, \ldots, q+p$, which requires pq transpositions. Thus (1.5.2) becomes

$$(g \wedge f)(x_1, \ldots, x_{p+q}) = (-1)^{pq} \sum_\sigma \epsilon(\sigma) g(x_{\sigma(p+1)}, \ldots, x_{\sigma(p+q)}) f(x_{\sigma(1)}, \ldots, x_{\sigma(p)})$$

the summation being over those σ satisfying (1.4.3). On the right-hand side each product may be permuted,

$$g(x_{\sigma(p+1)}, \ldots, x_{\sigma(p+q)}) \quad \text{with} \quad f(x_{\sigma(1)}, \ldots, x_{\sigma(p)}),$$

since multiplication of scalars is commutative. Thus we see that the right-hand side is also equal to

$$(-1)^{pq}(f \wedge g)(x_1, \ldots, x_{p+q}),$$

which proves (1.5.1).

PROPOSITION 1.5.2. *Exterior multiplication of multilinear alternating forms is associative.* In other words, if $f \in \mathscr{A}_p(\mathrm{E}; \mathbf{R})$, $g \in \mathscr{A}_q(\mathrm{E}; \mathbf{R})$, $h \in \mathscr{A}_r(\mathrm{E}; \mathbf{R})$ then

$$(f \wedge g) \wedge h = f \wedge (g \wedge h) \tag{1.5.4}$$

To prove this we need the following lemma.

Lemma 1.5.3. Let p, q, r be three integers >0. Let $\mathscr{A}_{p,q,r}(\mathrm{E}; \mathrm{F})$ be the subspace of $\mathscr{L}_{p+q+r}(\mathrm{E}; \mathrm{F})$ formed by those mappings which are alternating with respect to the first p variables, alternating with respect to the following q variables, and alternating with respect to the final r variables. Consider the diagram

$$\begin{array}{ccc} \mathscr{A}_{p,q,r}(\mathrm{E}; \mathrm{F}) & \xrightarrow{\varphi_{p,q}} & \mathscr{A}_{p+q,r}(\mathrm{E}; \mathrm{F}) \\ \downarrow{\scriptstyle \varphi_{q,r}} & & \downarrow{\scriptstyle \varphi_{p+q,r}} \\ \mathscr{A}_{p,q+r}(\mathrm{E}; \mathrm{F}) & \xrightarrow{\varphi_{p,q+r}} & \mathscr{A}_{p+q+r}(\mathrm{E}; \mathrm{F}). \end{array}$$

In this diagram $\varphi_{p,q}$ transforms $u \in \mathscr{A}_{p,q,r}(\mathrm{E}; \mathrm{F})$ into $\tilde{u}$, an alternation with respect to the first $p + q$ variables (without touching the remaining r), viz.,

$$\tilde{u}(x_1, \ldots, x_{p+q+r}) = \sum_\sigma \epsilon(\sigma) u(x_{\sigma(1)}, \ldots, x_{\sigma(p+q+r)}),$$

the summation being over permutations σ which leave $p + q + 1, \ldots, p + q + r$ fixed and which satisfy (1.4.3). We define $\varphi_{q,r}$ in the same way (for which the permutations σ leave $1, \ldots, p$ fixed). Then *the diagram is commutative,* i.e.,

$$\varphi_{p+q,r} \circ \varphi_{p,q} = \varphi_{p,q+r} \circ \varphi_{q,r}.$$

[This is the assertion of the lemma.]

PROOF OF THE LEMMA. We shall show that if $u \in \mathscr{A}_{p,q,r}(\mathrm{E}; \mathrm{F})$, then each of the alternating mappings

$$\varphi_{p+q,r}(\varphi_{p,q}(u)) \quad \text{and} \quad \varphi_{p,q+r}(\varphi_{q,r}(u))$$

is equal to the mapping v defined by

$$v(x_1, \ldots, x_{p+q+r}) = \sum_\sigma \epsilon(\sigma) u(x_{\sigma(1)}, \ldots, x_{\sigma(p+q+r)}),$$

where the summation is over those permutations of $\{1, \ldots, p + q + r\}$ satisfying

$$\begin{cases} \sigma(1) < \cdots < \sigma(p), \quad \sigma(p+1) < \cdots < \sigma(p+q) \\ \sigma(p+q+1) < \cdots < \sigma(p+q+r). \end{cases} \tag{1.5.5}$$

Now this is virtually intuitive if we think of a pack of cards which has been divided into three piles containing p, q, r cards respectively. The mapping $\varphi_{p,q}$ "shuffles" the first two piles; next $\varphi_{p+q,r}$ "shuffles" the combination thus obtained with the third pile. On the other hand, $\varphi_{q,r}$ "shuffles" the last two piles, then $\varphi_{p,q+r}$ "shuffles" the first pile with the combination obtained. In either case we obtain a "shuffling" of the three piles, expressed by (1.5.5) relative to the permutation σ finally obtained. We shall limit ourselves to this brief indication of the proof, leaving the details to the reader.

Let us now use Lemma 1.5.3 to prove Prop. 1.5.2. For the product $(f \wedge g) \wedge h$, consider the multilinear mapping

$$u(x_1, \ldots, x_{p+q+r}) = (f(x_1, \ldots, x_p)g(x_{p+1}, \ldots, x_{p+q}))h(x_{p+q+1}, \ldots, x_{p+q+r})$$

which belongs to $\mathscr{A}_{p,q,r}$, and apply to it the transformation $\varphi_{p+q,r} \circ \varphi_{p,q}$. For the product $f \wedge (g \wedge h)$ consider

$$u_1(x_1, \ldots, x_{p+q+r}) = f(x_1, \ldots, x_p)(g(x_{p+1}, \ldots, x_{p+q})h(x_{p+q+1}, \ldots, x_{p+q+r}))$$

and apply to it the transformation $\varphi_{p,q+r} \circ \varphi_{q,r}$. Now $u_1 = u$ because of *the associative law of scalar multiplication.* And since, by the lemma, $\varphi_{p+q,r} \circ \varphi_{p,q} = \varphi_{p,q+r} \circ \varphi_{q,r}$, the equality (1.5.4) is finally established.

1.6 *The exterior product of n linear forms*

Since exterior multiplication is associative we can consider finite products of the form $f_1 \wedge f_2 \wedge \cdots \wedge f_n$. Let us examine in particular the case in which $f_1, \ldots, f_n$ are elements of $\mathscr{A}_1(\mathrm{E}; \mathbf{R}) = \mathscr{L}(\mathrm{E}; \mathbf{R})$.

PROPOSITION 1.6.1. *If $f_1, \ldots, f_n$ are linear forms, then*

(1.6.1) $$f_1 \wedge \cdots \wedge f_n = \sum_{\sigma} \epsilon(\sigma) f_1(x_{\sigma(1)}) \cdots f_n(x_{\sigma(n)}),$$

the summation being over the n! permutations of $\{1, \ldots, n\}$.

PROOF. Use induction on n. The result is trivial for $n = 1$, and has already been established for $n = 2$ (relation (1.4.9)). The remainder of the proof is left as an exercise for the reader.

Let us consider the $n \times n$ matrix which has the element $f_i(x_j)$ at the intersection of the ith row and jth column. The right-hand side of (1.6.1) will be recognized as the value of the determinant of this matrix. Hence we have the *important result*

(1.6.2) $$(f_1 \wedge \cdots \wedge f_n)(x_1, \ldots, x_n) = \det \{f_i(x_j)\}$$

Observe that if $x_1, \ldots, x_n \in \mathrm{E}$ are fixed, both sides of (1.6.2) are n-linear alternating functions of $f_1, \ldots, f_n$. In fact $f_1 \wedge \cdots \wedge f_n$ vanishes whenever $f_i = f_{i+1}$, since we have already seen (§ 1.4) that $f \wedge f = 0$.

Exercise. In order that the n elements $f_1, \ldots, f_n \in \mathscr{L}(\mathrm{E}; \mathbf{R}) = \mathrm{E}^*$ be linearly dependent it is necessary and sufficient that $f_1 \wedge \cdots \wedge f_n = 0$. The main points of proof are:

1. If for example, $f_n = \sum_{i=1}^{n-1} \lambda_i f_i$, then $f_1 \wedge \cdots \wedge f_n = 0$.

2. If $f_1, \ldots, f_n$ are linearly independent, there exist $x_1, \ldots, x_n \in \mathrm{E}$ such that $f_i(x_j) = \delta_{ij}$ (Kronecker delta); in other words the matrix $\{f_i(x_j)\}$ is the unit matrix. Therefore its determinant is unity, and we have

$$(f_1 \wedge \cdots \wedge f_n)(x_1, \ldots, x_n) = 1,$$

which proves that the n-linear alternating form is not identically zero.

1.7 *The case where* E *is of finite dimension*

Suppose that E has dimension k, and choose a base for E. Then E is identified with $\mathbf{R}^k$. For $i = 1, \ldots, k$, we shall denote by $u_i: \mathbf{R}^k \to \mathbf{R}$ the ith coordinate form.

THEOREM 1.7.1. *Every p-linear alternating mapping* $f \in \mathscr{A}_p(\mathbf{R}^k; \mathrm{F})$ *possesses a unique description of the form*

$$f = \sum_{1 \leqslant i_1 \leqslant \cdots \leqslant i_p \leqslant k} c_{i_1, \ldots, i_p} u_{i_1} \wedge \cdots \wedge u_{i_p}; \tag{1.7.1}$$

where the constants $c_{i_1, \ldots, i_p}$ *are elements of* F.

On the right-hand side each term is a product of a constant $c_{i_1, \ldots, i_p} \in \mathrm{F}$ and a p-linear alternating form $u_{i_1} \wedge \cdots \wedge u_{i_p}$, an exterior product of p linear forms.

PROOF. First of all, f is a multilinear mapping (we recall that every multilinear mapping is continuous, since $\mathrm{E} = \mathbf{R}^k$ is of finite dimension). Referring to Chap. 1, § 6.2 of *Differential Calculus*, we see that

$$f(x^1, \ldots, x^p) = \sum_{i_1, \ldots, i_p} c_{i_1, \ldots, i_p} x^1_{i_1} \cdots x^p_{i_p}; \tag{1.7.2}$$

$x^1, \ldots, x^p$ denote p vectors; $x^i_1, \ldots, x^i_p$ denote the coordinates of the ith vector x^i; in the summation the integers $i_1, \ldots, i_p$ vary independently over $1, \ldots, k$. The "coefficients" $c_{i_1, \ldots, i_p} \in \mathrm{F}$ are uniquely determined by f. If we set

$$f(x^{\sigma(1)}, \ldots, x^{\sigma(p)}) = \epsilon(\sigma) f(x^1, \ldots, x^p),$$

then, because of the uniqueness of the coefficients, we find that

$$c_{i_{\sigma(1)}, \ldots, i_{\sigma(p)}} = \epsilon(\sigma) c_{i_1, \ldots, i_p}, \tag{1.7.3}$$

and in particular $c_{i_1, \ldots, i_p}$ vanishes each time that two of the integers $i_1, \ldots, i_p$ are equal. The conditions (1.7.3) are necessary and sufficient for the p-linear form having coefficients $c_{i_1, \ldots, i_p}$ to be *alternating*.

On the right-hand side of (1.7.2) let us group together the $p!$ terms which can be derived from each other by a permutation of the *distinct* integers $i_1, \ldots, i_p$. We see that

$$f(x^1, \ldots, x^p) = \sum_{i_1 < \cdots < i_p} c_{i_1, \ldots, i_p} \Big(\sum_{\sigma} \epsilon(\sigma) x^{\sigma(1)}_{i_1} \cdots x^{\sigma(p)}_{i_p} \Big);$$

the coefficients $c_{i_1, \ldots, i_p}$, for $i_1 < \cdots < i_p$, can be chosen arbitrarily. This is the most general p-linear *alternating* form. On the right-hand side,

$$\sum_{\sigma} \epsilon(\sigma) x^{\sigma(1)}_{i_1} \cdots x^{\sigma(p)}_{i_p}$$

by (1.6.1), is none other than the value for $x^1, \ldots, x^p$ of the exterior product $u_{i_1} \wedge \cdots \wedge u_{i_p}$ of the coordinate forms $u_{i_1}, \ldots, u_{i_p}$. This proves (1.7.1).

COROLLARY 1.7.2. If $p > k$, the vector space $\mathscr{A}_p(\mathbf{R}^k; \mathbf{R})$ is reduced to 0. (In fact, there does not exist a strictly increasing sequence $i_1 < \cdots < i_p$ consisting of p integers $\geqslant 1$ and $\leqslant k$.)

If $p = k$, every element of $\mathscr{A}_k(\mathbf{R}^k; \mathbf{R})$ is of the form

$$\boxed{c u_1 \wedge \cdots \wedge u_n} \quad \text{with} \quad c \in \mathrm{F}.$$

COROLLARY 1.7.3. If $F = \mathbf{R}$, the vector space $\mathscr{A}_p(\mathbf{R}^k; \mathbf{R})$ has a base, formed by the elements

$$u_{i_1} \wedge \cdots \wedge u_{i_p}$$

corresponding to all the *strictly increasing sequences* $i_1 < \cdots < i_p$ of integers $\geqslant 1$ and $\leqslant k$. This base is empty for $p > k$, and contains only one member for $p = k$.

Exercise. Calculate the dimension of the vector space $\mathscr{A}_p(\mathbf{R}^k; \mathbf{R})$ for $p < k$.

2. Differential forms

2.1 *Definition of a differential form*

In the following, U will denote an open subset of a Banach space E, and F a Banach space. We observe that $\mathscr{A}_p(E; F)$ (the space of p-linear continuous *alternating* mappings $E^p \to F$) is a Banach space, since it is a closed vector subspace of the Banach space $\mathscr{L}_p(E; F)$ (cf. § 1.1).

DEFINITION. A mapping

$$\omega: U \to \mathscr{A}_p(E; F) \tag{2.1.1}$$

is called a *differential form of degree p, defined in* U *and with values in* F. (We also say, more shortly, a *differential p-form* defined in U with values in F.) A differential form ω is said to be of class C^n if the mapping (2.1.1) is of class C^n; n is an integer $\geqslant 0$ and can be ∞.

Particular cases. A differential form of degree 0 is none other than a *function* $U \to F$. A differential form of degree 1 is a mapping $U \to \mathscr{L}(E; F)$.

Notation. $\Omega_p^{(n)}(U, F)$ will denote the set of all differential p-forms of class C^n, defined in U, with values in F. Evidently it is *a vector space.*

For $\omega \in \Omega_p^{(n)}(U, F)$, $x \in U$, and $\xi_1, \ldots, \xi_p \in E$,

$$\omega(x) \circ (\xi_1, \ldots, \xi_p) \in F$$

denotes the value of $\omega(x) \in \mathscr{A}_p(E; F)$ for the sequence of vectors $\xi_1, \ldots, \xi_p$. We also use the notation

$$\omega(x; \xi_1, \ldots, \xi_p).$$

For fixed x, this is a multilinear alternating function of $\xi_1, \ldots, \xi_p$.

Example. Let $f: U \to F$ be a mapping of class C^n with $n \geqslant 1$. Then the derived function $f': U \to \mathscr{L}(E; F)$ may be considered as a differential form of degree 1, of class C^{n-1}, defined in U, with values in F.

2.2 *Operations on differential forms*

In § 1 we defined an exterior product of multilinear alternating mappings. This now enables us to introduce an *exterior product for differential forms*. To be more explicit:

Let F, G, H be three Banach spaces, and let $\Phi: F \times G \to H$ be a linear continuous mapping. Now let

$$\alpha \in \Omega_p^{(n)}(U, F), \qquad \beta \in \Omega_q^{(n)}(U, G).$$

For each $x \in U$, $\alpha(x)$ is an element of $\mathscr{A}_p(E; F)$, and $\beta(x)$ an element of $\mathscr{A}_q(E; G)$. They have the exterior product

$$\alpha(x) \underset{\Phi}{\wedge} \beta(x) \in \mathscr{A}_{p+q}(E; H).$$

The mapping

$$x \mapsto \alpha(x) \underset{\Phi}{\wedge} \beta(x)$$

of U into $\mathscr{A}_{p+q}(E; H)$ is of class C^n, for the mapping is composed of

$$x \mapsto (\alpha(x), \beta(x))$$

which is of class C^n, and of the continuous mapping

$$\mathscr{A}_p(E; F) \times \mathscr{A}_q(E; G) \to \mathscr{A}_{p+q}(E; H)$$

defined by exterior multiplication.

DEFINITION. *The exterior product of the differential forms* α *and* β, denoted by $\alpha \underset{\Phi}{\wedge} \beta$, is the differential form

$$x \mapsto \alpha(x) \underset{\Phi}{\wedge} \beta(x).$$

If α and β are of class C^n, their exterior product $\alpha \underset{\Phi}{\wedge} \beta$ is of class C^n. We have thus defined a bilinear mapping

$$\Omega_p^{(n)}(U, F) \times \Omega_q^{(n)}(U, G) \to \Omega_{p+q}^{(n)}(U, H).$$

Using (1.4.6) the explicit form of this definition is

$$(2.2.1) \qquad (\alpha \underset{\Phi}{\wedge} \beta)(x; \xi_1, \ldots, \xi_{p+q}) = \sum_\sigma \epsilon(\sigma)\Phi(\alpha(x; \xi_{\sigma(1)}, \ldots, \xi_{\sigma(p)}), \beta(x; \xi_{\sigma(p+1)}, \ldots, \xi_{\sigma(p+q)})),$$

the summation being over all permutations σ of $\{1, \ldots, p+q\}$ satisfying

$$\sigma(1) < \cdots < \sigma(p), \qquad \sigma(p+1) < \cdots < \sigma(p+q).$$

Example. Let $f: U \to F$ be a function, and $\omega: U \to \mathscr{A}_n(E; \mathbf{R})$ a differential n-form with scalar values. Let us take, therefore, $G = \mathbf{R}$, $H = F$, Φ being the multiplication defined by the vectorial structure of F. Then the product $f \underset{\Phi}{\wedge} \omega$, which we write simply as $f \cdot \omega$ is the differential form defined by

$$(2.2.2) \qquad (f \cdot \omega)(x; \xi_1, \ldots, \xi_n) = f(x) \cdot \omega(x; \xi_1, \ldots, \xi_n).$$

Let us consider a second example, that in which α, β are differential forms of degree 1 taking scalar values. We take for $\Phi: \mathbf{R} \times \mathbf{R} \to \mathbf{R}$ scalar multiplication. Then $\alpha \underset{\Phi}{\wedge} \beta$, which we write more simply as $\alpha \wedge \beta$, is defined by the formula

$$(2.2.3) \qquad (\alpha \wedge \beta)(x; \xi_1, \xi_2) = \alpha(x; \xi_1)\beta(x; \xi_2) - \alpha(x; \xi_2)\beta(x; \xi_1).$$

Exterior multiplication of differential forms possesses the same properties as multiplication of multilinear alternating mappings.

PROPOSITION 2.2.1. Exterior multiplication of differential forms with scalar values

$$(\alpha, \beta) \mapsto \alpha \wedge \beta$$

is *anticommutative*: $\beta \wedge \alpha = (-1)^{pq}\alpha \wedge \beta$ if α is of degree p and β of degree q; it is also associative: $(\alpha \wedge \beta) \wedge \gamma = \alpha \wedge (\beta \wedge \gamma)$. This follows immediately from Prop. 1.5.1 and 1.5.2.

PROPOSITION 2.2.2. Let $\omega_1, \ldots, \omega_p \in \Omega_1^{(n)}(U, \mathbf{R})$ be differential 1-forms of class C^n with scalar values. Then $\omega_1 \wedge \cdots \wedge \omega_p \in \Omega_p^{(n)}(U, \mathbf{R})$ is a multilinear alternating function of $\omega_1, \ldots, \omega_p$; we have

$$(2.2.4) \qquad (\omega_1 \wedge \cdots \wedge \omega_n)(x; \xi_1, \ldots, \xi_n) = \sum_\sigma \epsilon(\sigma)\omega_1(x; \xi_{\sigma(1)}) \ldots \omega_n(x; \xi_{\sigma(n)})$$
$$= \det\{\omega_i(x; \xi_j)\}.$$

The proof follows directly from Prop. 1.6.1.

Remark. For a given integer n, consider the vector space $\bigoplus_{p \geqslant 0} \Omega_p^{(n)}(U, \mathbf{R})$, the direct sum over all values of p of the vector spaces $\Omega_p^{(n)}(U, \mathbf{R})$. The exterior product

$$\Omega_p^{(n)}(U, \mathbf{R}) \times \Omega_q^{(n)}(U, \mathbf{R}) \to \Omega_{p+q}^{(n)}(U, \mathbf{R})$$

extends linearly, and makes an *algebra* of

$$\Omega^{(n)}(U, \mathbf{R}) = \bigoplus_{p \geqslant 0} \Omega_p^{(n)}(U, \mathbf{R}).$$

It is called a *graduated algebra*, since the product of an element of degree p with an element of degree q is an element of degree $p + q$. It is *anticommutative* and *associative.*

2.3 *Exterior differentiation*

We now consider an important operation which has no analogue in the theory of alternating mappings. We shall associate with each $\omega \in \Omega_p^{(n)}(U, F)$ (with $n \geqslant 1$) a differential $(p + 1)$-form, denoted by $d\omega$:

$$d\omega \in \Omega_{p+1}^{(n-1)}(U, F),$$

and called the *exterior differential of* ω.

Suppose then that

$$\omega \to \mathscr{A}_p(E; F)$$

is a mapping of class C^n $(n \geqslant 1)$. Let us consider the derived mapping; it is of class C^{n-1}:

$$\omega' : U \to \mathscr{L}(E; \mathscr{A}_p(E; F));$$

for $x \in U$,

$$(\omega'(x) \cdot \xi_0) \cdot (\xi_1, \ldots, \xi_p) \in F$$

is a multilinear continuous function of $\xi_0, \ldots, \xi_p \in E$, and an *alternating* function of $\xi_1, \ldots, \xi_p$. In other words (with the notation of § 1.4), $\omega'(x)$ may be considered as an element of $\mathscr{A}_{1,p}(E; F)$. Now in § 1.4 we have defined a linear continuous mapping

$$\varphi_{1,p} : \mathscr{A}_{1,p}(E; F) \to \mathscr{A}_{p+1}(E; F).$$

DEFINITION. The exterior differential $d\omega$ is the composite mapping

$$U \xrightarrow{\omega'} \mathscr{A}_{1,p}(E; F) \xrightarrow{\varphi_{1,p}} \mathscr{A}_{p+1}(E; F).$$

Explicitly:

$$(2.3.1) \qquad \boxed{(d\omega)(x; \xi_0, \ldots, \xi_p) = \sum_{i=0}^{p} (-1)^i(\omega'(x)\cdot\xi_i)\cdot(\xi_0, \ldots, \hat{\xi}_i, \ldots, \xi_p)}$$

If ω is of class C^n, $d\omega$ is of class C^{n-1}.

EXAMPLES. 1. Take ω to be the function f: $U \to F$. Then

$$(df)(x; \xi) = f'(x)\cdot\xi.$$

Here, df: $U \to \mathscr{L}(E; F)$ is none other than the *derived* function f'.

2. For $p = 1$, we have

$$(2.3.2) \qquad (d\omega)(x; \xi_1, \xi_2) = (\omega'(x)\cdot\xi_1)\cdot\xi_2 - (\omega'(x)\cdot\xi_2)\cdot\xi_1.$$

From which we have:

PROPOSITION 2.3.1. Let $\omega \in \Omega_1^{(n)}(U, F)$, $n \geqslant 1$. In order that $d\omega = 0$ it is necessary and sufficient that, for all $x \in U$, the bilinear mapping

$$(\xi_1, \xi_2) \mapsto (\omega'(x)\cdot\xi_1)\cdot\xi_2$$

be *symmetric*.

2.4 *Properties of exterior differentiation*

PROPOSITION 2.4.1. *If f is a function of class C^1, and ω a differential p-form, then*

$$(2.4.1) \qquad d(f\cdot\omega) = (df) \wedge \omega + f\cdot(d\omega).$$

(Two particular cases are worthy of note: (1) f has values in F, ω is a scalar, $f\cdot\omega$ has values in F; (2) f is a scalar, ω has values in F, and $f\cdot\omega$ takes values in F.)

PROOF. In general, let f take values in F, ω take values in G, and let Φ: $F \times G \to H$ be a bilinear continuous mapping. Then $f\cdot\omega$ is the mapping

$$x \mapsto \Phi(f(x), \omega(x))$$

of U into $\mathscr{A}_p(E; H)$. Now it is known (cf. Chap. 1, (2.5.5) *Differential Calculus*) that the derivative of this function of x takes the value, for the vector $\xi \in E$,

$$\Phi(f'(x)\cdot\xi, \omega(x)) + \Phi(f(x), \omega'(x)\cdot\xi).$$

Therefore, limiting ourselves for simplicity to the case where Φ is $F \times R \to F$, or $R \times F \to F$, we can write:

$$(2.4.2) \qquad (f\cdot\omega)'(x)\cdot\xi = (f'(x)\cdot\xi)\cdot\omega(x) + f(x)\cdot(\omega'(x)\cdot\xi).$$

Using (2.3.1), with ω replaced by $f\cdot\omega$, we have:

$$d(f\cdot\omega)\cdot(x; \xi_0, \ldots, \xi_p) = \sum_{i=0}^{p} (-1)^i((f\cdot\omega)'(x)\cdot\xi_i)\cdot(\xi_0, \ldots, \hat{\xi}_i, \ldots, \xi_p),$$

which, taking account of (2.4.2), gives

$$\sum_{i=0}^{p} (-1)^i (f'(x)\cdot\xi_i)\cdot\omega(x)\cdot(\xi_0,\ldots,\hat{\xi}_i,\ldots,\xi_p)$$
$$+\sum_{i=0}^{p} (-1)^i f(x)\cdot(\omega'(x)\cdot\xi_i)\cdot(\xi_0,\ldots,\hat{\xi}_i,\ldots,\xi_p).$$

The first summation is the value of the exterior product $(df)\wedge\omega$ for $(\xi_0,\ldots,\xi_p)$; the second summation is the value of $f\cdot(d\omega)$ for $(\xi_0,\ldots,\xi_p)$. This proves (2.4.1).

THEOREM 2.4.2. *Let* $\alpha\in\Omega_p^{(n)}(\mathrm{U},\mathbf{R})$ *and* $\beta\in\Omega_q^{(n)}(\mathrm{U},\mathbf{R})$ *be two differential forms of class* $\mathrm{C}^n(n\geqslant 1)$. *Then*

(2.4.3) $$d(\alpha\wedge\beta)=(d\alpha)\wedge\beta+(-1)^p\alpha\wedge(d\beta)$$

PROOF. By the formula which gives the derivative of a bilinear function of two functions (Formula (2.5.4), Chap. 1, *Differential Calculus*), we have, for $x\in\mathrm{U}$ and $\xi\in\mathrm{E}$,

$$(\alpha\wedge\beta)'(x)\cdot\xi=(\alpha'(x)\cdot\xi)\wedge\beta(x)+\alpha(x)\wedge(\beta'(\alpha)\cdot\xi).$$

(This is an equality between elements of $\mathscr{A}_{p+q}(\mathrm{E};\mathrm{F})$.)

For simplicity let us write $\alpha',\beta,\alpha,\beta',(\alpha\wedge\beta)'$ instead of $\alpha'(x),\beta(x)$, etc...., remembering that we are concerned with the values of these functions at the point x. Thus we have,

(2.4.4) $$(\alpha\wedge\beta)'\cdot\xi=(\alpha'\cdot\xi)\wedge\beta+\alpha\wedge(\beta'\cdot\xi),$$

or again, since exterior multiplication is anticommutative:

(2.4.5) $$(\alpha\wedge\beta)'\cdot\xi=(\alpha'\cdot\xi)\wedge\beta+(-1)^{pq}(\beta'\cdot\xi)\wedge\alpha.$$

Put

$$u(\xi_0,\xi_1,\ldots,\xi_{p+q})=((\alpha'\cdot\xi_0)\wedge\beta)\circ(\xi_1,\ldots,\xi_{p+q})$$
$$v(\xi_0,\xi_1,\ldots,\xi_{p+q})=((\beta'\cdot\xi_0)\wedge\alpha)\circ(\xi_1,\ldots,\xi_{p+q})$$
$$w(\xi_0,\xi_1,\ldots,\xi_{p+q})=((\alpha\wedge\beta)'\cdot\xi_0)\circ(\xi_1,\ldots,\xi_{p+q}).$$

For fixed x these are elements of $\mathscr{A}_{1,p+q}(\mathrm{E};\mathbf{R})$, and (2.4.5) becomes

$$w=u+(-1)^{pq}v.$$

We are attempting to calculate the $(p+q+1)$-linear alternating function $d(\alpha\wedge\beta)$, which is, by definition, the transformation of w by means of the canonical mapping

$$\varphi_{1,p+q}:\mathscr{A}_{1,p+q}(\mathrm{E};\mathbf{R})\to\mathscr{A}_{p+q+1}(\mathrm{E};\mathbf{R}).$$

Thus we have

(2.4.6) $$d(\alpha\wedge\beta)=\varphi_{1,p+q}(u)+(-1)^{pq}\varphi_{1,p+q}(v).$$

Now u is an exterior product; therefore it is obtained by applying to

$$((\alpha'\cdot\xi_0)\cdot(\xi_1,\ldots,\xi_p))\cdot\beta(\xi_{p+1},\ldots,\xi_{p+q}),$$

considered as an element of $\mathscr{A}_{p,q}(\mathrm{E};\mathbf{R})$ (for fixed ξ_0), the mapping

$$\varphi_{p,q}:\mathscr{A}_{p,q}(\mathrm{E};\mathbf{R})\to\mathscr{A}_{p+q}(\mathrm{E};\mathbf{R}).$$

Finally, $\varphi_{1,p+q}(u)$ is transformed by $\varphi_{1,p+q} \circ \varphi_{p,q}$ from the element of $\mathscr{A}_{1,p,q}(E; \mathbf{R})$ defined by

$$(2.4.7) \qquad (\xi_0, \xi_1, \ldots, \xi_{p+q}) \mapsto ((\alpha' \cdot \xi_0) \cdot (\xi_1, \ldots, \xi_p)) \cdot \beta(\xi_{p+1}, \ldots, \xi_{p+q}).$$

Now by Lemma (1.5.3) we have

$$\varphi_{1,p+q} \circ \varphi_{p,q} = \varphi_{1+p,q} \circ \varphi_{1,p}.$$

If we apply $\varphi_{1,p}$ to the multilinear form (2.4.7) we obtain

$$(\xi_0, \xi_1, \ldots, \xi_{p+q}) \mapsto ((d\alpha) \cdot (\xi_0, \xi_1, \ldots, \xi_p)) \cdot \beta(\xi_{p+1}, \ldots, \xi_{p+q}),$$

and if we now apply $\varphi_{1+p,q}$ we obtain the value of the exterior product $d\alpha \wedge \beta$ on the system of vectors $(\xi_0, \ldots, \xi_{p+q})$.

Thus, finally we have

$$\varphi_{1,p+q}(u) = (d\alpha) \wedge \beta.$$

Similarly, we have

$$\varphi_{1,p+q}(v) = (d\beta) \wedge \alpha.$$

Thus (2.4.6) gives

$$d(\alpha \wedge \beta) = (d\alpha) \wedge \beta + (-1)^{pq}(d\beta) \wedge \alpha.$$

Now exterior multiplication is anticommutative, hence

$$(d\beta) \wedge \alpha = (-1)^{p(q+1)}\alpha \wedge (d\beta),$$

and our final result is

$$d(\alpha \wedge \beta) = (d\alpha) \wedge \beta + (-1)^{p}\alpha \wedge (d\beta),$$

which is the relation (2.4.3) to be established.

2.5 *Fundamental property of exterior differentiation*

THEOREM 2.5.1. *Let* $\omega \in \Omega_p^{(n)}(U, F)$ *be a differential form of class* C^n *with* $n \geqslant 2$. *Then*

$$\boxed{d(d\omega) = 0}.$$

(The operation d, applied twice, gives zero.)

PROOF. For a given $x \in U$ (henceforth taken as read in the notation), $d\omega$ is a multilinear alternating function of $\xi_1, \ldots, \xi_{p+1}$, that is obtained by applying $\varphi_{1,p}: \mathscr{A}_{1,p}(E; F) \to \mathscr{A}_{p+1}(E; F)$ to the multilinear mapping

$$(\xi_1, \ldots, \xi_{p+1}) \mapsto \omega'(\xi_1) \cdot (\xi_2, \ldots, \xi_{p+1}).$$

Thus, for fixed ξ_0, $(d\omega)' \cdot \xi_0$ is a multilinear alternating function of $\xi_1, \ldots, \xi_{p+q}$ obtained by applying $\varphi_{1,p}$ to the multilinear mapping

$$(\xi_1, \ldots, \xi_{p+1}) \mapsto ((\omega'' \cdot \xi_0) \cdot \xi_1) \cdot (\xi_2, \ldots, \xi_{p+1}),$$

which we shall also write as

$$(2.5.1) \qquad \omega''(\xi_0, \xi_1) \cdot (\xi_2, \ldots, \xi_{p+1})$$

considering ω'' to be a bilinear function with values in $\mathscr{A}_p(E; F)$. Let us consider (2.5.1) as a multilinear function of $\xi_0, \xi_1, \ldots, \xi_{p+1}$; it is an element of $\mathscr{A}_{1,1,p}(E; F)$. Also the operation $\varphi_{1,p}$, applied to (2.5.1) considered (for fixed ξ_0) as a function of $\xi_1, \ldots, \xi_{p+1}$, gives an element of $\mathscr{A}_{1,p+1}(E; F)$ which is none other than $(d\omega)'$ considered as a multilinear function of $\xi_0, \xi_1, \ldots, \xi_{p+1}$. Next one obtains the $(p + 2)$-linear alternating function $d(d\omega)$ by applying $\varphi_{1,p+1}$ to $(d\omega)'$, i.e., by applying $\varphi_{1,p+1} \circ \varphi_{1,p}$ to the multilinear function (2.5.1).

By Lemma 1.5.3 we have

$$\varphi_{1,p+1} \circ \varphi_{1,p} = \varphi_{2,p} \circ \varphi_{1,1}.$$

Now $\varphi_{1,1}$, applied to the multilinear function (2.5.1), gives

$$\omega''(\xi_0, \xi_1)\cdot(\xi_2, \ldots, \xi_{p+1}) - \omega''(\xi_1, \xi_0)\cdot(\xi_2, \ldots, \xi_{p+1}),$$

and *this is zero* since the second derivative ω'' is a linear *symmetric* function of ξ_0 and ξ_1 (Chap. 1, Theorem 5.1.1, *Differential Calculus*). Hence $d(d\omega) = 0$.

2.6 *Differential forms on a finite dimensional space*

Suppose E is of finite dimension k. The choice of a base for E identifies E with $\mathbf{R}^k$. Let $u_i \in \mathscr{L}(\mathbf{R}^k; \mathbf{R})$ be the ith coordinate function, considered as a linear form. Given an open set $U \subset \mathbf{R}^k$, let x_i denote the restriction of u_i to U considered this time as a differentiable mapping $x_i \colon U \to \mathbf{R}$.

Lemma 2.6.1. The differential dx_i of the function x_i is the *constant* mapping $U \to \mathscr{L}(\mathbf{R}^k; \mathbf{R})$ whose value is $u_i \in \mathscr{L}(\mathbf{R}^k; \mathbf{R})$.

Indeed, the function x_i is linear; by Prop. 2.4.2 of Chap. 1 of *Differential Calculus* we know that the derived mapping is constant, and that the value of this constant is the element of $\mathscr{L}(\mathbf{R}^k; \mathbf{R})$ precisely equal to the function considered.

THEOREM 2.6.2. *If* $U \subset \mathbf{R}$ *is open, every differential form* $\omega \in \Omega_p^{(n)}(U, F)$ *may be written uniquely in the form*

$$\omega = \sum_{i_1 < \cdots < i_p} c_{i_1 \ldots i_p}(x)\, dx_{i_1} \wedge \cdots \wedge dx_{i_p}, \tag{2.6.1}$$

where the "coefficients" $c_{i_1, \ldots, i_p}$ are functions of class $C^n \colon U \to F$. In the summation the integers $i_1 < \cdots < i_p$, are $\geqslant 1$ and $\leqslant k$.

PROOF. Let us refer to Theorem 1.7.1 which furnishes a canonical representation of the elements of $\mathscr{A}_p(\mathbf{R}^k; F)$. Here, ω is, by definition, a mapping of class C^n:

$$\omega \colon U \to \mathscr{A}_p(\mathbf{R}^k; F).$$

For $x \in U$, $\omega(x) \in \mathscr{A}_p(\mathbf{R}^k; F)$ may be written uniquely in the form

$$\sum_{i_1 < \cdots < i_p} c_{i_1 \ldots i_p} u_{i_1} \wedge \cdots \wedge u_{i_p};$$

now (by the preceding lemma) $u_{i_1}, \ldots, u_{i_p}$ are the values, for $x \in U$, of the constant functions $dx_{i_1}, \ldots, dx_{i_p}$; the "coefficients" $c_{i_1, \ldots, i_p}$ depend on x. It is evident that the function

$$x \mapsto \sum_{i_1 < \cdots < i_p} c_{i_1 \ldots i_p}(x) u_{i_1} \wedge \cdots \wedge u_{i_p}$$

is of class C^n if, and only if, the $c_{i_1, \ldots, i_p}$ are functions of class C^n. Bearing in mind the definition of the exterior multiplication of the differential forms which appear on the right in (2.6.1), this proves Theorem 2.6.2.

In future the right-hand side of (2.6.1) will be referred to as the *canonical form* of the differential form ω (in the open set $U \subset \mathbf{R}^k$).

Theorem 2.6.2 is valid in the particular case where $F = \mathbf{R}$ (differential forms with scalar values); in this case the coefficients $c_{i_1, \ldots, i_p}$ are numerical functions. We can speak of $\Omega_p^{(n)}(U, \mathbf{R})$ as a *module on the ring* $\Omega_0^{(n)}(U, \mathbf{R})$ of numerical functions of class C^n, this module having for its *base* the set of elements

$$dx_{i_1} \wedge \cdots \wedge dx_{i_p} \in \Omega_p^{(n)}(U; \mathbf{R})$$

relative to the increasing sequence $i_1 < \cdots < i_p$ of integers $\geqslant 1$, $\leqslant k$.

Particular case: $p = 1$. In this case every differential 1-form can be written uniquely in the form

$$\omega = \sum_{i=1}^{k} c_i(x)\, dx_i \tag{2.6.2}$$

where the c_i are mappings of class $C^n: U \to F$. We recognize (2.6.2) as the usual form for differential forms of degree 1.

It is important to be able to calculate the value $\omega(x; \xi)$ of the form ω defined by (2.6.2) for a point $x \in U$ and a vector $\xi \in \mathbf{R}^k$. To do this, recall that, by Lemma 2.6.1, the value of dx_i for $(x; \xi)$ is independent of x, and equal to $u_i(\xi)$: the value of the ith coordinate form u_i on the vector ξ. Thus

$$\omega(x; \xi) = \sum_{i=1}^{p} c_i(x)\, \xi_i \tag{2.6.3}$$

ξ_i being the ith coordinate of the vector $\xi \in \mathbf{R}^k$.

PROPOSITION 2.6.3. Let U be an open subset of $\mathbf{R}^k$. If $f: U \to F$ is of class C^1, the differential df has the canonical form

$$\boxed{df = \sum_{i=1}^{k} \frac{\partial f}{\partial x_i}\, dx_i}. \tag{2.6.4}$$

Indeed, $df: U \to \mathscr{L}(\mathbf{R}^k; F)$ is none other than the derived mapping f'; and we have seen in Chap. 1, Formula (2.6.1) of *Differential Calculus*, that if ξ is a vector with coordinates $\xi_1, \ldots, \xi_k$, then

$$f'(x) \cdot \xi = \sum_{i=1}^{k} \frac{\partial f}{\partial x_i}(x)\, \xi_i.$$

Thus:

$$df(x; \xi) = \sum_{i=1}^{k} \frac{\partial f}{\partial x_i}(x)\, \xi_i,$$

and by (2.6.3) we obtain precisely the relation (2.6.4).

2.7 *Calculus of operations on canonically represented differential forms*

In this section we shall suppose that U is an open subset of $\mathbf{R}^k$ and write our differential forms in the canonical form of Theorem 2.6.2. Let α be a p-form and β a q-form; we propose to determine the canonical form of $\alpha \wedge \beta$ given the canonical representations of α and of β.

Since exterior multiplication is distributive with respect to addition, it suffices to do the calculation for the case where α and β each consists of only one term:

$$\alpha = a(x)\, dx_{i_1} \wedge \cdots \wedge dx_{i_p},$$
$$\beta = b(x)\, dx_{j_1} \wedge \cdots \wedge dx_{j_q}.$$

We require the product $\alpha \wedge \beta$ relative to a bilinear continuous mapping

$$\Phi: \mathrm{F} \times \mathrm{G} \to \mathrm{H}$$

(F denoting the space of values of α, i.e., of the function $a(x)$, and G denoting the space of values of β, i.e., of the function $b(x)$). Using the associativity and anticommutativity of multiplication, we find

$$\alpha \underset{\Phi}{\wedge} \beta = \Phi(a(x), b(x))\, dx_{i_1} \wedge \cdots \wedge dx_{i_p} \wedge dx_{j_1} \wedge \cdots \wedge dx_{j_q}.$$

Here $\Phi(a, b)$ is the "product" of the functions a and b relative to Φ (for example, if a and b are numerical functions, and if $\Phi: \mathbf{R} \times \mathbf{R} \to \mathbf{R}$ is a multiplication of scalars, $\Phi(a, b)$ is simply the product of the functions a and b). It remains to determine the *canonical form of the product*

$$dx_{i_1} \wedge \cdots \wedge dx_{i_p} \wedge dx_{j_1} \wedge \cdots \wedge dx_{j_q}.$$

Let us distinguish *two cases*: (1) if the integers $i_1, \ldots, i_p, j_1, \ldots, j_q$, are not distinct, the product is zero; (2) if the integers are all distinct; in this case a permutation σ renders them in a strictly increasing sequence

$$k_1 < k_2 < \cdots < k_{p+q};$$

and we have

$$dx_{i_1} \wedge \cdots \wedge dx_{i_p} \wedge dx_{j_1} \wedge \cdots \wedge dx_{j_q} = \epsilon(\sigma)\, dx_{k_1} \wedge \cdots \wedge dx_{k_{p+q}}.$$

We obtain thus the canonical form of $\alpha \underset{\Phi}{\wedge} \beta$.

We have just seen how to calculate the exterior product of canonical forms. Let us now see how to calculate to *exterior derivative* $d\omega$ of a differential form ω written in canonical form. It is sufficient to do the calculation for ω consisting of one term only:

(2.7.1) $$\omega = c(x)\, dx_{i_1} \wedge \cdots \wedge dx_{i_p},$$

the "coefficient" c being a mapping $\mathrm{U} \to \mathrm{F}$ of class C^1. We shall apply the results of § 2.4 for the differential $d(\alpha \wedge \beta)$; by repeated application of the result we see that if ω is an exterior product

$$\omega = \omega_1 \wedge \cdots \wedge \omega_n$$

of differential forms ω of degree p, then

(2.7.2) $$d\omega = d\omega_1 \wedge \omega_2 \wedge \cdots \wedge \omega_n + (-1)^{p_1}\omega_1 \wedge (d\omega_2) \wedge \cdots \wedge \omega_n + \cdots + (-1)^{p_1+p_2+\cdots+p_{n-1}}\omega_1 \wedge \cdots \wedge \omega_{n-1} \wedge (d\omega_n).$$

Apply this to the form ω defined by (2.7.1):

$$d\omega = dc \wedge dx_{i_1} \wedge \cdots \wedge dx_{i_p} + c\, d(dx_{i_1}) \wedge dx_{i_2} \wedge \cdots \wedge dx_{i_p} + \cdots \pm c\, dx_{i_1} \wedge \cdots \wedge dx_{i_{p-1}} \wedge d(dx_{i_p}).$$

Now, by § 2.5, we have

$$d(dx_{i_1}) = 0, \ldots, d(dx_{i_p}) = 0.$$

Thus we finally obtain:

$$(2.7.3) \qquad \boxed{d\omega = dc \wedge dx_{i_1} \wedge \cdots \wedge dx_{i_p}},$$

a *formula which is simple and easy* (and important!) *to remember.*

(2.7.3) is not the canonical form of $d\omega$ (for ω defined by (2.7.1)). But this is easily obtained. Indeed on the right-hand side of (2.7.3), let us replace dc by its value

$$\sum_{j=1}^{k} \frac{\partial c}{\partial x_j} dx_j \qquad \text{(cf. Prop. 2.6.3)},$$

to obtain

$$d\omega = \sum_{j=1}^{k} \frac{\partial c}{\partial x_j} dx_j \wedge dx_{i_1} \wedge \cdots \wedge dx_{i_p}.$$

It remains to find the canonical form of $dx_j \wedge dx_{i_1} \wedge \cdots \wedge dx_{i_p}$; as we have seen above, this is 0 if j is equal to one of the integers $i_1, \ldots, i_p$, and equal to $\epsilon(\sigma)\, dx_{k_1} \wedge \cdots \wedge dx_{k_{p+1}}$, if σ is a permutation of $(j, i_1, \ldots, i_p)$ transforming it into a strictly increasing sequence $(k_1, \ldots, k_{p+1})$.

Example. In $\mathbf{R}^3$ with three coordinates x, y, z, consider

$$\omega = P\, dx + Q\, dy + R\, dz,$$

the coefficients P, Q, R being functions of class C^1 of x, y, z. A simple calculation gives the canonical form of

$$d\omega = dP \wedge dx + dQ \wedge dy + dR \wedge dz;$$

viz.,

$$d\omega = \left(\frac{\partial R}{\partial y} - \frac{\partial Q}{\partial z}\right) dy \wedge dz + \left(\frac{\partial P}{\partial z} - \frac{\partial R}{\partial x}\right) dz \wedge dx + \left(\frac{\partial Q}{\partial x} - \frac{\partial P}{\partial y}\right) dx \wedge dy.$$

Exercise. Let us verify, by writing in canonical form, that $d(d\omega) = 0$ for all differential forms ω of class C^2 (cf. Theorem 2.5.1). We start by proving that $d(df) = 0$ for a function f of class C^2; by Prop. (2.6.3)

$$df = \sum_j \frac{\partial f}{\partial x_j} dx_j,$$

from which,

$$d(df) = \sum_j d\left(\frac{\partial f}{\partial x_j}\right) dx_j.$$

Now

$$d\left(\frac{\partial f}{\partial x_j}\right) = \sum_i \frac{\partial^2 f}{\partial x_i \partial x_j} dx_i,$$

giving

$$d(df) = \sum_{i,j} \frac{\partial^2 f}{\partial x_i \partial x_j} dx_i \wedge dx_j;$$

the indices i, j vary independently in the summation. Terms with $i = j$ are zero, since $dx_i \wedge dx_i = 0$. For $i \neq j$, group together the (i, j) term and the (j, i) term, to obtain

$$d(df) = \sum_{i<j} \left(\frac{\partial^2 f}{\partial x_i \partial x_j} dx_i \wedge dx_j + \frac{\partial^2 f}{\partial x_j dx_i} dx_j \wedge dx_i \right).$$

Now (cf. Schwarz's theorem, Prop. 5.2.2, Chap. 1, *Differential Calculus*)

$$\frac{\partial^2 f}{\partial x_i \partial x_j} = \frac{\partial^2 f}{\partial x_j \partial x_i},$$

and also $dx_i \wedge dx_j = -dx_j \wedge dx_i$. Thus we indeed find

$$d(df) = 0.$$

To verify now that $d(d\omega) = 0$, it suffices to consider the case where ω consists of one term only,

$$\omega = c \, dx_{i_1} \wedge \cdots \wedge dx_{i_p}.$$

Now,

$$d\omega = (dc) \wedge dx_{i_1} \wedge \cdots \wedge dx_{i_p}.$$

By equation (2.7.2) (with ω replaced by $d\omega$), $d(d\omega)$ consists of a sum of terms each of which contains a factor $d(dc)$, or $d(dx_{i_1}), \ldots,$ or $d(dx_{i_p})$. But, by the above, each of these factors is zero. Thus $d(d\omega) = 0$. Hence, the use of the canonical form has furnished us with a new proof (valid for an open subset of $\mathbf{R}^k$) of Theorem 2.5.1: $d(d\omega) = 0$.

2.8 *Change of variable*

Let us recall the general case where U is an open subset of a Banach space E. Let

$$\omega: \mathrm{U} \to \mathscr{A}_p(\mathrm{E}; \mathrm{F})$$

be a differential p-form of class C^n. Suppose further that we are given a mapping

$$\varphi: \mathrm{U}' \to \mathrm{U},$$

where U′ is an open subset of the Banach space E′; let φ be of class C^{n+1}. Let us define a p-form of class C^n:

$$\mathrm{U}' \to \mathscr{A}_p(\mathrm{E}'; \mathrm{F}),$$

denoted by $\varphi^*(\omega)$, or simply $\varphi^*\omega$, which we shall call *the differential form obtained from* ω *by the change of variable*

$$\varphi: \mathrm{U}' \to \mathrm{U}.$$

Before giving a precise definition of the form $\varphi^*\omega$, let us note that, for a point $y \in \mathrm{U}'$ and for the vectors $\eta_1, \ldots, \eta_p \in \mathrm{E}'$, we wish its value to be given by the formula

$$(\varphi^*\omega)(y; \eta_1, \ldots, \eta_p) = \omega(\varphi(y); \varphi'(y) \cdot \eta_1, \ldots, \varphi'(y) \cdot \eta_p). \tag{2.8.1}$$

(Recall that $\varphi'(y)$ is an element of $\mathscr{L}(E'; E)$.) It remains to show that (2.8.1) does indeed define $\varphi^*\omega$ as a mapping of class C^n of U' into $\mathscr{A}_p(E'; F)$.

Let us begin with the case $p = 0$; ω is then simply the function $f: U \to F$ of class C^n; (2.8.1) is thus

$$(\varphi^* f)(y) = f(\varphi(y)),$$

in other words $\varphi^* f: U' \to F$ is, *by definition*, the *composite function* $f \circ \varphi$. In this case ($p = 0$), it suffices to suppose that φ is of class C^n in order to conclude that $f \circ \varphi$ is of class C^n (cf. Chap. 1, Theorem 5.4.5 of *Differential Calculus*). Thus the mapping

$$f \mapsto f \circ \varphi = \varphi^*(f)$$

is a linear mapping $\Omega_0^{(n)}(U, F) \to \Omega_0^{(n)}(U', F)$.

Now consider the case where $p > 0$; this time it will be necessary to suppose that φ is of class C^{n+1}. Equation (2.8.1) shows that $\varphi^*\omega$, considered as the mapping $U' \to \mathscr{A}_p(E'; F)$, is composed of the two following mappings.

(1) The mapping $U' \to \mathscr{A}_p(E; F) \times \mathscr{L}(E'; E)$, of which the first component is $\omega \circ \varphi: U' \to \mathscr{A}_p(E'; F)$, and the second component is $\varphi': U' \to \mathscr{L}(E'; E)$.

(2) The mapping $\lambda_p: \mathscr{A}_p(E; F) \times \mathscr{L}(E'; E) \to \mathscr{A}_p(E'; F)$ which, for $f \in \mathscr{A}_p(E; F)$ and $g \in \mathscr{L}(E'; E)$, associates the element of $\mathscr{A}_p(E'; F)$ defined by

$$(\eta_1, \ldots, \eta_p) \mapsto f(g(\eta_1), \ldots, g(\eta_p)).$$

It is easy to see that the mapping λ_p is of class C^∞ (it is even a polynomial mapping, for it is linear in f, and a homogeneous polynomial of degree p in g). Since ω is supposed to be of class C^n, and φ of class C^{n+1}, the mappings $\omega \circ \varphi$ and φ' are of class C^n; hence the mapping (1) is of class C^n. Thus the mapping $\varphi^*\omega$, composed of (1) and (2) is of class C^n.

Q.E.D.

We have thus proved:

PROPOSITION 2.8.1. *If φ is of class C^{n+1}, the "change of variable" defined by equation* (2.8.1) *associates with each differential form $\omega \in \Omega_p^{(n)}(U, F)$ a differential form $\varphi^*\omega \in \Omega_p^{(n)}(U', F)$.* Thus φ^* is a linear mapping

$$\varphi^* = \Omega_p^{(n)}(U; F) \to \Omega_p^{(n)}(U', F).$$

(For $p = 0$ it suffices to suppose that φ is of class C^n.)

2.9 *Properties of the change of variable mapping φ^**

THEOREM 2.9.1. *φ^* preserves exterior products:* if $\alpha \in \Omega_p^{(n)}(U, F)$ and $\beta \in \Omega_q^{(n)}(U, G)$, and if the bilinear continuous mapping $\Phi: F \times G \to H$ is given, then

$$\varphi^*(\alpha \underset{\Phi}{\wedge} \beta) = (\varphi^*\alpha) \underset{\Phi}{\wedge} (\varphi^*\beta). \tag{2.9.1}$$

PROOF. The result is verified by a simple calculation, using the definition of exterior multiplication. Writing, for simplicity, φ' instead of $\varphi'(y)$, we have

$$(\varphi^*\alpha \underset{\Phi}{\wedge} \varphi^*\beta)(y; \eta_1, \ldots, \eta_{p+q}) \tag{2.9.2}$$
$$= \sum_\sigma \epsilon(\sigma)\Phi((\varphi^*\alpha)(y; \eta_{\sigma(1)}, \ldots, \eta_{\sigma(p)}), (\varphi^*\beta)(y; \eta_{\sigma(p+1)}, \ldots, \eta_{\sigma(p+q)})),$$

the summation being taken over permutations σ of $\{1, \ldots, p + q\}$ which satisfy (1.4.3). This is equal to

$$\sum_{\sigma} \epsilon(\sigma)\Phi(\alpha(\varphi(y): \varphi' \cdot \eta_{\sigma(1)}, \ldots, \varphi' \cdot \eta_{\sigma(p)}), \beta(\varphi(y); \varphi' \cdot \eta_{\sigma(p+1)}, \ldots, \varphi' \cdot \eta_{\sigma(p+q)})).$$

This is clearly equal to $(\varphi^*\omega)(y; \eta_1, \ldots, \eta_{p+q})$, where ω denotes the differential form $U \to \mathscr{A}_{p+q}(E; F)$ defined by

$$\omega(x; \xi_1, \ldots, \xi_{p+q}) = \sum_{\sigma} \epsilon(\sigma)\Phi(\alpha(x; \xi_{\sigma(1)}, \ldots, \xi_{\sigma(p)}), \beta(x; \xi_{\sigma(p+1)}, \ldots, \xi_{\sigma(p+q)})).$$

We see that

$$\omega = \alpha \underset{\Phi}{\wedge} \beta;$$

thus the left-hand side of (2.9.2) is equal to the value of $\varphi^*(\alpha \underset{\Phi}{\wedge} \beta)$ at the point y, for the vectors $\eta_1, \ldots, \eta_{p+q} \in E'$. This establishes (2.9.1).

THEOREM 2.9.2. *If* $\varphi: U' \to U$ *and* $f: U \to F$ *are of class* C^1, *then*

(2.9.3) $$\varphi^*(df) = d(\varphi^*f)$$

(note that $\varphi^*f: U' \to F$ is of class C^1). *If* $\varphi: U' \to U$ *is of class* C^2, *and*

$$\omega: U \to \mathscr{A}_p(E: F)$$

is of class C^1, *then*

(2.9.4) $$\varphi^*(d\omega) = d(\varphi^*\omega)$$

(note that $\varphi^*\omega$ and $d\omega$ are of class C^1).

In brief, φ^* is compatible with exterior differentiation.

PROOF of (2.9.3). We have

(2.9.5) $$(df)(x; \xi) = f'(x) \cdot \xi,$$

hence

$$(\varphi^*(df))(y; \eta) = f'(\varphi(y)) \cdot \varphi'(y) \cdot \eta = (f'(\varphi(y)) \circ \varphi'(y)) \cdot \eta.$$

Now (derivative of a composite function):

$$f'(\varphi(y)) \circ \varphi'(y) = (f \circ \varphi)'(y) = (\varphi^*f)'(y),$$

hence the members of (2.9.5) are equal to

$$(\varphi^*f)'(y) \cdot \eta = (d(\varphi^*f))(y; \eta),$$

which proves the equality (2.9.3).

The proof of (2.9.4) is a little more complicated, but it is still a matter of simple verification by calculation. This will be left as an exercise; in the particular case $E = \mathbf{R}^k$ we shall present a proof which utilizes the canonical form of differential forms.

2.10 *Calculation of* φ^* *in canonical form*

We know that every $\omega \in \Omega_p^{(n)}(U, F)$, U an open subset of $\mathbf{R}^k$, is a sum of differential forms of the type

$$c(x)\, dx_{i_1} \wedge \cdots \wedge dx_{i_p}.$$

Let us calculate $\varphi^*\omega$ in the case where $\omega = c(x)\, dx_{i_1} \wedge \cdots \wedge dx_{i_p}$. By Theorem 2.9.1 (compatibility of φ^* with the exterior product) and the relation (2.9.3), we have

$$\varphi^*\omega = \varphi^*(c)\, d(\varphi^* x_{i_1}) \wedge \cdots \wedge d(\varphi^* x_{i_p}).$$

Now $\varphi^* x_i = x_i \circ \varphi : \mathrm{U}' \to \mathbf{R}$ is the function φ_i which expresses the ith coordinate of $\varphi(y)$ as a function of $y \in \mathrm{U}$. Thus

$$\varphi^*\omega = c(\varphi(y))(d\varphi_{i_1}) \wedge \cdots \wedge (d\varphi_{i_p}). \tag{2.10.1}$$

To obtain the canonical form of $\varphi^*\omega$, it remains to use the explicit form

$$d\varphi_i = \sum_j \frac{\partial \varphi_i}{\partial y_j}\, dy_j$$

in the right-hand side of (2.10.1), and then to develop it as explained in § 2.7.

Working rule. To effect the transformation φ, express the coordinates $x_1, \ldots, x_k$ of $\varphi(y)$ as functions of the coordinates $y_1, \ldots, y_h$ of y (having supposed that U′ is an open subset of $\mathbf{R}^h$):

$$x_i = \varphi_i(y_1, \ldots, y_h);$$

next, in $c(x)\, dx_{i_1} \wedge \cdots \wedge dx_{i_p}$, replace x by $\varphi(y)$, dx_{i_1} by $d\varphi_{i_1}$, etc. . . .; finally develop the differentials $d\varphi_{i_1}, \ldots$ as linear combinations of $dy_1, \ldots, dy_h$.

We now propose to establish the truth of (2.9.4). We shall do this for

$$\omega = c(x)\, dx_{i_1} \wedge \cdots \wedge dx_{i_p}.$$

We have

$$\begin{aligned} \varphi^*(d\omega) &= \varphi^*(dc \wedge dx_{i_1} \wedge \cdots \wedge dx_{i_p}) \\ &= d(c \circ \varphi) \wedge d\varphi_{i_1} \wedge \cdots \wedge d\varphi_{i_p} \\ &= d((c \circ \varphi)\, d\varphi_{i_1} \wedge \cdots \wedge d\varphi_{i_p}) \\ &= d(\varphi^*\omega). \end{aligned}$$

Exercise. Let us suppose that $k = h = p$. A differential form ω of degree p in $\mathrm{U} \subset \mathbf{R}^p$ is written

$$\omega = c(x)\, dx_1 \wedge \cdots \wedge dx_p$$

(there is only one term). The mapping $\varphi : \mathrm{U}' \to \mathrm{U}$ is defined by p functions of p variables:

$$x_i = \varphi_i(y_1, \ldots, y_p).$$

Verify that

$$dx_1 \wedge \cdots \wedge dx_p = J\, dy_1 \wedge \cdots \wedge dy_p,$$

where J is the *Jacobian* of the transformation φ:

$$J = \det\left(\left\{\frac{\partial \varphi_i}{\partial x_j}\right\}\right),$$

that is also denoted by

$$\frac{\partial(x_1, \ldots, x_p)}{\partial(y_1, \ldots, y_p)}.$$

Thus we have

$$\boxed{\varphi^*\omega = c(\varphi(y))\,\frac{\partial(x_1,\ldots,x_p)}{\partial(y_1,\ldots,y_p)}\,dy_1 \wedge\cdots\wedge dy_p}. \tag{2.10.2}$$

2.11 *Transitivity of the change of variable*

Let E, E′, E″ be Banach spaces; U, U′, U″ open subsets of E, E′, E″ respectively. Let

$$\varphi\colon \mathrm{U}' \to \mathrm{U},\ \psi\colon \mathrm{U}'' \to \mathrm{U}',$$

be mappings of class C^{n+1}. We know that $\varphi \circ \psi\colon \mathrm{U}'' \to \mathrm{U}$ is of class C^{n+1}. The change of variable defines the linear mappings

$$\varphi^*\colon \Omega_p^{(n)}(\mathrm{U}, \mathrm{F}) \to \Omega_p^{(n)}(\mathrm{U}', \mathrm{F})$$
$$\psi^*\colon \Omega_p^{(n)}(\mathrm{U}', \mathrm{F}) \to \Omega_p^{(n)}(\mathrm{U}'', \mathrm{F}).$$

PROPOSITION 2.11.1. *The mapping*

$$(\varphi \circ \psi)^*\colon \Omega_p^{(n)}(\mathrm{U}, \mathrm{F}) \to \Omega_p^{(n)}(\mathrm{U}'', \mathrm{F})$$

is given in terms of φ^* *and* ψ^* *by*

$$\boxed{(\varphi \circ \psi)^* = \psi^* \circ \varphi^*}. \tag{2.11.1}$$

(*N.B.* the symbols φ and ψ are not in the same order on each side of this equality.)

PROOF. Let us begin with the case $p = 0$; it then suffices to suppose that φ and ψ are of class C^n, and the relation (2.11.1) expresses that, for every $f\colon \mathrm{U} \to \mathrm{F}$ of class C^n, we have

$$f \circ (\varphi \circ \psi) = (f \circ \varphi) \circ \psi,$$

which is evidently the case. In the general case let $\omega \in \Omega_p^{(n)}(\mathrm{U}, \mathrm{F})$; then, for $y \in \mathrm{U}'$, $\eta_1, \ldots, \eta_p \in \mathrm{E}'$:

$$(\varphi^*\omega)(\eta; \eta_1, \ldots, \eta_p) = \omega(\varphi(y); \varphi'(y)\cdot\eta_1, \ldots, \varphi'(y)\cdot\eta_p).$$

Hence, for $z \in \mathrm{U}''$, $\xi_1, \ldots, \xi_p \in \mathrm{E}''$, we have

$$(\psi^*(\varphi^*\omega))(z; \xi_1, \ldots, \xi_p) = \omega(\varphi(\psi(z)); \varphi'(\psi(z)\cdot\psi'(z)\cdot\xi_1, \ldots)). \tag{2.11.2}$$

Now

$$\varphi'(\psi(z))\cdot\psi'(z) = (\varphi \circ \psi)'(z),$$

by the theorem which gives the derivative of a function of a function. The right-hand side of (2.11.2) is therefore equal to

$$\omega((\varphi \circ \psi)(z); (\varphi \circ \psi)'(z)\cdot\xi_1, \ldots, (\varphi \circ \psi)'(z)\cdot\xi_p),$$

and this proves (2.11.1).

2.12 *Condition for a differential form to equal dα*

Suppose we are given a differential form

$$\omega: \mathrm{U} \to \mathscr{A}_p(\mathrm{E}; \mathrm{F}) \quad \text{with} \quad p \geqslant 1.$$

Under what condition does there exist a differential form $\alpha: \mathrm{U} \to \mathscr{A}_{p-1}(\mathrm{E}; \mathrm{F})$ such that $d\alpha = \omega$?

If we demand that α be of class C^2, then it is *necessary* that ω be of class C^1 and that $d\omega = 0$, since $d(d\alpha) = 0$. We shall give a partial converse to this result when the open set U satisfies certain conditions (see below).

DEFINITION. A subset U of a vector space E is called *starlike* with respect to one of its points a, if, for every $x \in \mathrm{U}$, the segment $[a, x]$, consisting of the points $(1 - t)a + tx$ (where $0 \leqslant t \leqslant 1$), is contained in U. It is evident that a starlike set is *connected*, and that a *convex* set is starlike.

THEOREM 2.12.1 (Poincaré's theorem). *Let* E *and* F *be two Banach spaces, and let* U *be an open subset of* E, *starlike with respect to one of its points. If*

$$\omega \in \Omega_p^{(n)}(\mathrm{U}, \mathrm{F}) \qquad (n \geqslant 1, p \geqslant 1)$$

satisfies $d\omega = 0$, *then there exists* $\alpha \in \Omega_{p-1}^{(n)}(\mathrm{U}, \mathrm{F})$ *such that* $d\alpha = \omega$.

The proof of this result will be given later. For the present recall the case $p = 0$: if a function $f \in \Omega_0^1(\mathrm{U}, \mathrm{F})$ satisfies $df = 0$, then f is constant in U, at least if U is connected (in particular if U is starlike).

We shall be using a lemma which is independent of the theory of differential forms:

Lemma 2.12.2 (differentiation under the integral sign).

Let E and F be two Banach spaces, I the segment [0, 1], and let

$$\varphi: \mathrm{U} \times \mathrm{I} \to \mathrm{F}$$

be a continuous mapping. For $x \in \mathrm{U}$, set

$$\psi(x) = \int_0^1 \varphi(x, t)\, dt.$$

Then $\psi: \mathrm{U} \to \mathrm{F}$ is continuous. If, further, the partial derivative φ'_x exists at every point $(x, t) \in \mathrm{U} \times \mathrm{I}$ and is a continuous mapping $\mathrm{U} \times \mathrm{I} \to \mathscr{L}(\mathrm{E}; \mathrm{F})$, then ψ is of class C^1, and

$$\psi'(x) = \int_0^1 \varphi'_x(x, t)\, dt. \tag{2.12.1}$$

PROOF. Choose $\epsilon > 0$. For each $(x, t) \in \mathrm{U} \times \mathrm{I}$ there exists $\eta(x, t)$ such that

$$\|\varphi(x', t') - \varphi(x, t)\| \leqslant \tfrac{1}{2}\epsilon \quad \text{for} \quad \|x' - x\| \leqslant \eta(x, t), |t' - t| \leqslant \eta(x, t).$$

Thus, in particular, we have

$$\|\varphi(x, t') - \varphi(x, t)\| \leqslant \tfrac{1}{2}\epsilon \quad \text{for} \quad |t' - t| \leqslant \eta(x, t),$$

so that, combining these inequalities, we have

$$\|\varphi(x', t') - \varphi(x, t')\| \leqslant \epsilon \quad \text{for} \quad \|x' - x\| \leqslant \eta(x, t), |t' - t| \leqslant \eta(x, t).$$

2 +

For each $t \in I$ let us associate the open interval of points t' which satisfy

$$|t' - t| < \eta(x, t);$$

(x being fixed). Since I is compact it can be covered by a *finite* number of such intervals; let the finite number of points $t_i (i \in I)$ be associated with these intervals. Denote the smallest member of $\eta(x, t_i)$ by $\eta(x)$. For every $t' \in I$ there exists t_i such that $|t' - t| < \eta(x, t_i)$; therefore

$$\|\varphi(x', t') - \varphi(x, t')\| \leqslant \epsilon \quad \text{for} \quad \|x' - x\| \leqslant \eta(x), \tag{2.12.2}$$

for every $t' \in I$. In other words, φ is continuous at the point x, *uniformly with respect to t.* Thus,

$$\psi(x') - \psi(x) = \int_0^1 (\varphi(x', t) - \varphi(x, t))\, dt;$$

so that

$$\|\psi(x') - \psi(x)\| \leqslant \int_0^1 \|\varphi(x', t) - \varphi(x, t)\|\, dt,$$

since the norm of the integral of a function is not greater than the integral of the norm of the function. Thus (2.12.2) shows that for $\|x' - x\| \leqslant \eta(x)$,

$$\|\psi(x') - \psi(x)\| \leqslant \int_0^1 \epsilon\, dt = \epsilon.$$

And this proves that $\psi: U \to F$ is a *continuous* function.

Now suppose that φ'_x exists and is a *continuous* mapping $U \times I \to \mathscr{L}(E; F)$. By what we have just proved, the function

$$\lambda(x) = \int_0^1 \varphi'_x(x, t)\, dt$$

is continuous. It remains to prove that ψ is differentiable, and that $\psi'(x) = \lambda(x)$. To do this we must show that, for x fixed,

$$\|\psi(x + h) - \psi(x) - \lambda(x)\cdot h\| = o(\|h\|). \tag{2.12.3}$$

Now, since φ' is *continuous*, given $\epsilon > 0$ there exists $\eta > 0$ (depending on x) such that

$$\|\varphi'_x(x + h, t) - \varphi'_x(x, t)\| \leqslant \epsilon \quad \text{provided that} \quad \|h\| \leqslant \eta,$$

for every $t \in I$; indeed, this is the result (2.12.2) proved for φ applied to φ'_x. Hence

$$\|\varphi(x + h, t) - \varphi(x, t) - \varphi'_x(x, t)\cdot h\| \leqslant \epsilon\|h\|$$

for $\|h\| \leqslant \eta$. Thus, we have, for $\|h\| \leqslant \eta$,

$$\|\psi(x + h) - \psi(x) - \lambda(x)\cdot h\| = \left\| \int_0^1 (\varphi(x + h, t) - \varphi(x, t) - \varphi'_x(x, t)\cdot h)\, dt \right\|$$

$$\leqslant \int_0^1 \|\varphi(x + h, t) - \varphi(x, t) - \varphi'_x(x, t)\cdot h\|\, dt \leqslant \int_0^1 \epsilon\|h\|\, dt = \epsilon\|h\|.$$

This proves (2.12.3), and completes the proof of the lemma.

COROLLARY 2.12.3. Under the hypotheses of the preceding lemma, if the nth partial derivative $\partial^n\varphi/\partial x^n$ exists and is a continuous function $U \times I \to \mathscr{L}_n(E; F)$, then ψ is of class C^n, and

$$\psi^{(n)}(x) = \int_0^1 \frac{\partial^n\varphi}{\partial x^n}(x, t)\, dt.$$

(The proof follows simply by induction on n.)

2.13 *Proof of Poincaré's theorem*

We shall suppose that U is starlike with respect to the origin; this can always be achieved by a translation.

We shall first establish Theorem 2.12.1 for the case $p = 1 (n \geqslant 1)$. Let

$$\omega \in \Omega_1^{(n)}(U, F)$$

be a 1-form of class C^n. Set

(2.13.1)
$$\boxed{f(x) = \int_0^1 \omega(tx; x)\, dt}.$$

Let us first show that f: $U \to F$ is a mapping of class C^n. Put $\varphi(x, t) = \omega(tx; x)$. After Lemma 2.12.2 and the Corollary 2.12.3, it is sufficient to show that, for t fixed, φ is of class C^n and that $\varphi, \varphi'_x, \ldots, \partial^n\varphi/\partial x^n$ are continuous mappings of $U \times I$ into F, $\mathscr{L}(E; F), \ldots, \mathscr{L}_n(E; F)$. To do this it suffices to show that φ: $U \times I \to F$ is of class C^n. Now φ is the composition of the following mappings:

(1) $U \times I \to U \times E$, which takes (x, t) into (tx, x); this mapping exists because U is starlike with respect to 0; it is of class C^∞.

(2) $U \times E \to \mathscr{L}(E; F) \times E$, which takes (x, ξ) into $(\omega(x), \xi)$; like ω, this mapping is of class C^n.

(3) $\mathscr{L}(E; F) \times E \to F$, which takes (f, ξ) into $f(\xi)$; this is a bilinear continuous mapping, and therefore of class C^∞.

Thus φ, composed of three mappings of class C^n is of class C^n. And consequently, the function f defined by (2.13.1) is of class C^n.

PROPOSITION 2.13.1. *In the preceding notation, suppose that* $d\omega = 0$. *Then* $df = \omega$. (This proves Poincaré's theorem for $p = 1$.)

PROOF. By Lemma 2.12.2,

$$f'(x) = \int_0^1 \varphi'_x(x, t)\, dt,$$

where $\varphi(x, t) = \omega(xt; x) = \omega(tx)\cdot x$. Let us calculate φ'_x; we have to find the derivative of a function of a function. Thus, for $\xi \in E$:

$$\varphi'_x \cdot \xi = t(\omega'_x(tx)\cdot\xi)\cdot x + \omega(tx)\cdot\xi.$$

Now the hypothesis $d\omega = 0$ expresses the fact that $(\omega'_x \cdot \xi_1) \cdot \xi_2$ is a bilinear *symmetric* function of ξ_1 and ξ_2. Hence

$$\varphi'_x \cdot \xi = t(\omega'_x(tx) \cdot x) \cdot \xi + \omega(tx) \cdot \xi,$$

i.e.,

$$\varphi'_x = t\omega'_x(tx) \cdot x + \omega(tx).$$

The right-hand side is equal to the derivative with respect to t of

$$t\omega(tx),$$

as is shown by simple calculation. Thus

$$f'(x) = \int_0^1 \frac{d}{dt}(t\omega(tx))\, dt$$

is equal to $g(1) - g(0)$, where $g(t) = t\omega(tx)$. Hence

$$f'(x) = \omega(x),$$

which proves that ω is equal to the differential form df.

Example. Suppose we are in $E = \mathbf{R}^n$. In canonical form we have

$$\omega = \sum_{i=1}^{n} c_i(x)\, dx_i,$$

the $c_i : U \to F$ being functions of class C^n. If $d\omega = 0$ we may write

$$\frac{\partial c_i}{\partial x_j} = \frac{\partial c_j}{\partial x_i} \quad \text{for} \quad 1 \leqslant i, j \leqslant n.$$

In this case (2.13.1) becomes:

$$f(x) = \sum_{i=1}^{n} \int_0^1 x_i c_i(tx)\, dt.$$

This function is of class C^n and satisfies $df = \omega$, i.e.,

$$\frac{\partial f}{\partial x_i} = c_i \quad \text{for} \quad 1 \leqslant i \leqslant n.$$

We now prove *Poincaré's theorem in the general case* (p arbitrary). We shall define, for every p, a linear mapping

$$k : \Omega_p^{(n)}(U, F) \to \Omega_{p-1}^{(n)}(U, F)$$

as follows: for $p = 0$ we take $k(f) = 0$ for all $f \in \Omega_0^{(n)}(U, F)$, which is the same as saying that the vector space $\Omega_{-1}^{(n)}(U, F)$ is reduced to zero. For $p \geqslant 1$, let $\omega \in \Omega_p^{(n)}(U, F)$; we shall define $k(\omega)$: this will be the differential form defined by

$$(2.13.2) \qquad \boxed{\alpha(x; \xi_1, \ldots, \xi_{p-1}) = \int_0^1 t^{p-1}\omega(tx; x, \xi_1, \ldots, \xi_p)\, dt}.$$

Naturally we must verify that $\alpha : U \to \mathcal{A}_{n-1}(E; F)$ is a well-defined differential ($p-1$)-

form of class C^n. To do this we proceed as in the case $p = 1$; let us denote by $\omega(tx, x)$ the $(p - 1)$-linear alternating mapping

$$(\xi_1, \ldots, \xi_{p-1}) \to \omega(tx; x, \xi_1, \ldots, \xi_{p-1}).$$

The right-hand side of (2.13.2), considered as a multilinear alternating function of $\xi_1, \ldots, \xi_{p-1}$, is

$$\int_0^1 t^{p-1}\omega(tx; x)\, dt;$$

i.e., $\alpha(x) \in \mathscr{A}_{p-1}(E; F)$. To show that α is of class C^n it is sufficient, after Lemma 2.12.2 and Corollary 2.12.3, to show that the mapping

$$(x, t) \to t^{p-1}\omega(tx; x)$$

is of class C^n. Now this is composed of the three following mappings:

(1) The mapping $U \times I \to U \times I \times E$ which transforms (x, t) into (tx, t, x); this mapping exists because U is starlike with respect to the origin; it is of class C^∞.

(2) The mapping $U \times I \times E \to \mathscr{A}_p(E; F) \times E$, which takes (x, t, ξ) into $(t^{p-1}\omega(x), \xi)$; like ω it is of class C^n.

(3) The mapping $\mathscr{A}_p(E; F) \times E \to \mathscr{A}_{p-1}(E; F)$, which transforms (f, ξ) into the element of $\mathscr{A}_{p-1}(E; F)$ defined by

$$(\xi_1, \ldots, \xi_{p-1}) \mapsto f(\xi, \xi_1, \ldots, \xi_{p-1}).$$

This mapping is of class C^∞.

Thus α, defined by (2.13.2), is a well-defined differential $(p - 1)$-form of class C^n.

PROPOSITION 2.13.2. *Under the preceding hypotheses* (U *starlike*): (a) if $\omega \in \Omega_p^{(n)}(U, F)$, with $n \geqslant 1, p \geqslant 1$, then

$$\boxed{d(k(\omega)) + k(d\omega) = \omega}; \tag{2.13.3}$$

(b) if $f \in \Omega_0^{(n)}(U, F)$, with $n \geqslant 1$, then

$$\boxed{k(df) = f - f_0}, \tag{2.13.4}$$

where f_0 is the constant function $U \to F$ whose value is $f(0)$.

Before proving this proposition, let us show that it includes, as a special case, Poincaré's theorem (Theorem 2.12.1): if, in (a), we suppose that $d\omega = 0$, relation (2.13.3) shows that

$$\omega = d(k(\omega)).$$

Thus the operator k furnishes a form $k(\omega) = \alpha$ such that $d\alpha = \omega$. Note also that (b) proves, again, that if $df = 0$, then f is constant.

PROOF OF PROPOSITION 2.13.2. First we shall verify (b): indeed, the function $k(df)$ is the function

$$x \mapsto \int_0^1 (f'(tx)\cdot x)\, dt = \int_0^1 \frac{d}{dt}(f(tx)) = f(x) - f(0).$$

The remainder of the proof may be left aside by the reader who dislikes calculation. However, it is given for the courageous reader! Let us therefore commence with (a), in which it is supposed that $p \geqslant 1$. Recall that

$$(2.13.5)\qquad (d\omega)(x;\xi_0,\xi_1,\ldots,\xi_p) = (\omega'(x)\cdot\xi_0)\cdot(\xi_1,\ldots,\xi_p) + \sum_{i=1}^{p}(-1)^i(\omega'(x)\cdot\xi_i)\cdot(\xi_0,\ldots,\hat{\xi}_i,\ldots,\xi_p).$$

Hence,

$$(d\omega)(tx;x,\xi_1,\ldots,\xi_p) = (\omega'(tx)\cdot x)\cdot(\xi_1,\ldots,\xi_p) + \sum_{i=1}^{p}(-1)^i(\omega'(tx)\cdot\xi_i)\cdot(x,\xi_1,\ldots,\hat{\xi}_i,\ldots,\xi_p).$$

Thus the form $k(d\omega) = \beta$ is defined by

$$(2.13.6)\qquad \beta(x:\xi_1,\ldots,\xi_p) = \int_0^1 t^p(\omega'(tx)\cdot x)\cdot(\xi_1,\ldots,\xi_p)\,dt + \sum_{i=1}^{p}(-1)^i\int_0^1 t^p(\omega'(tx)\cdot\xi_i)\cdot(x,\xi_1,\ldots,\hat{\xi}_i,\ldots,\xi_p)\,dt.$$

Furthermore, the differential $(p-1)$-form $k(\omega) = \alpha$ is given by (2.13.2); differentiating under the integral sign we obtain:

$$(2.13.7)\qquad (\alpha'(x)\cdot\xi_1)\cdot(\xi_2,\ldots,\xi_p) = \int_0^1 t^p(\omega'(tx)\cdot\xi_1)\cdot(x,\xi_2,\ldots,\xi_p)\,dt + \int_0^1 t^{p-1}\omega(tx;\xi_1,\xi_2,\ldots,\xi_p)\,dt.$$

Hence

$$(2.13.8)\qquad (d\alpha)(x;\xi_1,\ldots,\xi_p) = \sum_{i=1}^{p}(-1)^{i-1}\int_0^1 t^p(\omega'(tx)\cdot\xi_i)\cdot(x,\xi_1,\ldots,\hat{\xi}_i,\ldots,\xi_p)\,dt + \sum_{i=1}^{p}(-1)^{i-1}\int_0^1 t^{p-1}\omega(tx;\xi_i,\xi_1,\ldots,\hat{\xi}_i,\ldots,\xi_p)\,dt.$$

The second term on the right-hand side, taking account of the fact that ω is an alternating function of $\xi_1,\ldots,\xi_p$, is equal to

$$p\int_0^1 t^{p-1}\omega(tx;\xi_1,\ldots,\xi_p)\,dt.$$

We are now in a position to calculate the left-hand side of (2.13.3): the differential form $d(k(\omega)) + k(d\omega)$ is equal, in our usual notation, to $d\alpha + \beta$; the explicit form is obtained by adding equations (2.13.6) and (2.13.8). After simplification, we obtain:

$$\beta(x;\xi_1,\ldots,\xi_p) + d\alpha(x;\xi_1,\ldots,\xi_p) = \int_0^1 t^p(\omega'(tx)\cdot x)\cdot(\xi_1,\ldots,\xi_p)\,dt + p\int_0^1 t^{p-1}\omega(tx;\xi_1,\ldots,\xi_p)\,dt,$$

which may be put more simply in the form:

$$\beta(x) + d\alpha(x) = \int_0^1 [t^p(\omega'(tx)\cdot x) + pt^{p-1}\omega(tx)]\,dt \tag{2.13.9}$$

(equality between elements of $\mathscr{A}_p(E; F)$). Now

$$t^p(\omega'(tx)\cdot x) + pt^{p-1}\omega(tx) = \frac{d}{dt}(t^p\omega(tx)),$$

thus the right-hand side of (2.13.9) is equal to the difference between the values of $t^p\omega(tx)$ for $t = 1$ and $t = 0$, that is, finally, to $\omega(x)$. This completes the proof of (2.13.3).

We have now proved Prop. 2.13.2 and, at the same time, Poincaré's theorem.

Example. Suppose we are in $E = \mathbf{R}^3$, and that $F = \mathbf{R}$; let

$$\alpha = P\,dx + Q\,dy + R\,dz$$

be a differential form of class $C^n (n \geqslant 1)$: P, Q, R are therefore of class C^n and defined in an open $U \subset \mathbf{R}^n$, with numerical values. Then

$$d\alpha = \left(\frac{\partial R}{\partial y} - \frac{\partial Q}{\partial z}\right) dy \wedge dz + \left(\frac{\partial P}{\partial z} - \frac{\partial R}{\partial x}\right) dz \wedge dx + \left(\frac{\partial Q}{\partial x} - \frac{\partial P}{\partial y}\right) dx \wedge dy.$$

If the functions P, Q, R define a "vector field" in U (i.e., a mapping $U \to \mathbf{R}^3$), the field defined by

$$\frac{\partial R}{\partial y} - \frac{\partial Q}{\partial z}, \quad \frac{\partial P}{\partial z} - \frac{\partial R}{\partial x}, \quad \frac{\partial Q}{\partial x} - \frac{\partial P}{\partial y}$$

is called the *curl* of the field (P, Q, R). Conversely, given a 2-form

$$\omega = A\,dy \wedge dz + B\,dz \wedge dx + C\,dx \wedge dy$$

of class C^n, under what conditions does there exist a field (P, Q, R) of class C^n whose curl is equal to the vector field (A, B, C)? The condition is that there should exist a 1-form α of class C^n such that $d\alpha = \omega$. A *necessary* condition is that $d\omega = 0$ (at least if $n \geqslant 2$); Poincaré's theorem tells us that it is *sufficient* when U is *starlike.* The explicit form of this condition is easily obtained, since

$$d\omega = \left(\frac{\partial A}{\partial x} + \frac{\partial B}{\partial y} + \frac{\partial C}{\partial z}\right) dx \wedge dy \wedge dz,$$

hence the condition is that the "*divergence*" of the field (A, B, C), which by definition is the function

$$\frac{\partial A}{\partial x} + \frac{\partial B}{\partial y} + \frac{\partial C}{\partial z},$$

should be identically zero.

3. Curvilinear integral of a differential form of degree one

Let U denote an open subset of a Banach space E (the cases of interest will be those where $E = \mathbf{R}^n$ or $E = \mathbf{C}^n$).

3.1 *Paths of class* C^1

We have already defined (*Differential Calculus*, Chap. 1, § 3.3) a path of a topological space X. In the present case X = U, and a path of U is therefore a continuous mapping

$$\gamma\colon [a, b] \to U,$$

where $[a, b]$ denotes a compact interval $a \leqslant t \leqslant b$ of $\mathbf{R}$. (In § 3.3 the interval of t was taken to be $[0, 1]$.)

DEFINITION. A path γ is said to be of class C^1 if the mapping γ is of class C^1, i.e., if γ has a derivative $\gamma'(t)$ depending continuously on $t \in [a, b]$.

(*Remark.* Strictly speaking, the notion of mapping of class C^1 has only been defined for an open subset of a Banach space; in the present case we must therefore consider mappings defined in an *open interval* of $\mathbf{R}$. For a *closed* interval $[a, b]$ the derivative $\gamma'(a)$ is defined to be the derivative on the right at the point a, and $\gamma'(b)$ is the derivative on the left at b. It is easy to see that if $\gamma\colon [a, b] \to U$ is of class C^1 in this new sense, there exists an *open* interval I containing $[a, b]$, and a mapping $\gamma_1\colon I \to U$ of class C^1 of which γ is the restriction.)

DEFINITION. A path $\gamma\colon [a, b] \to U$ is said to be *piecewise of class* C^1 if there exists a finite subdivision of $[a, b]$:

$$a = t_0 < t_1 < \cdots < t_n = b \tag{3.1.1}$$

such that γ is of class C^1 when restricted to each of the segments $[t_i, t_{i+1}]\,(0 \leqslant i < n)$.

If this is the case, the derivative on the left $\gamma_l'(t_i)$ and the derivative on the right $\gamma_r'(t_i)$ (for $0 < i < n$) exist, but are not necessarily equal. Of course, if γ is a path which is piecewise of class C^1, there can exist several subdivisions of $[a, b]$ possessing the preceding property. If one of these is (3.1.1), and a second is

$$a = t_0' < t_1' < \cdots < t_p' = b, \tag{3.1.2}$$

then a third is obtained by *superposition*:

$$a = t_0'' < t_1'' < \cdots < t_q'' = b. \tag{3.1.3}$$

This is defined as follows: the set $(t_0'', \ldots, t_q'')$ is the union of the sets $(t_0, \ldots, t_n)$ and $(t_0', \ldots, t_p')$. (Thus $q \leqslant n + p - 1$.)

3.2 *The curvilinear integral*

Let ω be a differential form in U, of degree 1 and of class C^0, and with values in F; in other words, ω is a continuous mapping:

$$\omega\colon U \to \mathscr{L}(E; F).$$

As before, for $x \in U$ we shall denote by $\omega(x) \cdot \xi$ or $\omega(x; \xi)$ the value of $\omega(x) \in \mathscr{L}(E; F)$ for $\xi \in E$; it is an element of the Banach space F.

DEFINITION OF THE INTEGRAL $\int_\gamma \omega$ when γ is a path of class C^1 in U:

$$\gamma\colon [a, b] \to U.$$

By definition we have

$$(3.2.1) \qquad \int_\gamma \omega = \int_a^b f(t)\,dt,$$

where $f(t)\,dt$ is the differential form $\gamma^*(\omega)$: this is a differential form on $[a, b]$, obtained from ω by the change of variable γ of class C^1. The function $t \mapsto f(t)$ is continuous, and the explicit form of the change of variable is (cf. § 2.8)

$$(3.2.2) \qquad f(t) = \omega(\gamma(t)) \cdot \gamma'(t)$$

(where $\gamma'(t)$ is considered as an element of E, the space containing U). The right-hand side of (3.2.1) consists simply of the integral (in the Cauchy–Riemann sense) of a *continuous* function on a compact interval.

The classical properties of the integral apply to the present case: for a given path γ (of class C^1), $\int_\gamma \omega$ is a *linear* function of ω. Further, for a subdivision (3.1.1) of $[a, b]$, let γ_i denote the path corresponding to the segment $[t_i, t_{i+1}]$ $(0 \leqslant i < n)$, then

$$(3.2.3) \qquad \int_\gamma \omega = \sum_{i=0}^{n-1} \int_{\gamma_i} \omega,$$

for the simple reason that

$$\int_0^b f(t)\,dt = \sum_{i=0}^{n-1} \int_{t_i}^{t_{i+1}} f(t)\,dt.$$

We shall now define $\int_\gamma \omega$ when γ is a path piecewise of class C^1. Choose a subdivision (3.1.1) such that the restriction γ_i of γ to the segment $[t_i, t_{i+1}]$ is of class C^1, with i satisfying $0 \leqslant i \leqslant n - 1$. *By definition*, we put

$$\int_\gamma \omega = \sum_i \int_{\gamma_i} \omega,$$

the right-hand side being already well defined. To justify this definition it remains to verify: (1) that the value of the right-hand side is *independent of the choice of subdivision*; (2) that in the particular case in which γ is of class C^1, the new definition of $\int_\gamma \omega$ coincides with that already given above.

It is clear that if (1) is true, then (2) follows simply from equation (3.2.3), already established when γ is of class C^1. Let us prove (1): choose a second subdivision (3.1.2) such that the restriction of γ to each sement $[t'_j, t'_{j+1}]$ is of class C^1 $(0 \leqslant j < p)$. Suppose now that (3.1.3) is the subdivision obtained by superposition. Let $\bar{\gamma}_j$ denote the restriction of γ to $[t'_j, t'_{j+1}]$, and $\bar{\bar{\gamma}}_k$ the restriction of γ to $[t''_k, t''_{k+1}]$. Consider the sum

$$\sum_{k=0}^{q-1} \int_{\bar{\bar{\gamma}}_k} \omega;$$

it is clear that by the grouping together of certain terms we obtain

$$\sum_{i=0}^{n-1} \int_{\gamma_i} \omega,$$

and by grouping in a different manner

$$\sum_{j=0}^{p-1} \int_{\bar{\gamma}_j} \omega.$$

Thus we have

$$\sum_{i=0}^{n-1} \int_{\gamma_i} \omega = \sum_{j=0}^{p-1} \int_{\bar{\gamma}_j} \omega,$$

which justifies the definition given above.

Remark. Under the preceding hypotheses, we have

$$\gamma_i^*(\omega) = f_i(t)\, dt,$$

where f_i is a continuous function in $[t_i, t_{i+1}]$. The collection of functions f_i defines a function f in $[a, b]$, except perhaps at the points t_i $(1 \leqslant i \leqslant n-1)$ where f ceases to be continuous. Briefly, we say that f is a *piecewise continuous* function, and $\int_\gamma \omega$ is equal to the integral $\int_a^b f(t)\, dt$ of the piecewise continuous function f. With this convention, we can agree that, when γ is piecewise of class C^1, $\gamma^*(\omega)$ is equal to $f(t)\, dt$, where f is a piecewise continuous function (whose values at the points of discontinuity are not prescribed). Then equation (3.2.1), which defines $\int_\gamma \omega$ when γ is of class C^1, is also true when γ is piecewise of class C^1.

3.3 *Change of parameter*

Let $\varphi: [a', b'] \to [a, b]$ be a C^1-*diffeomorphism* of a segment $[a', b']$ onto a segment $[a, b]$ (cf. *Differential Calculus*, Chap. 1, § 4.1). The derivative $\varphi'(u)$ (for $a' \leqslant u \leqslant b'$) is non-zero; it therefore has a constant sign. Two cases are possible:

(1) $\varphi'(u) > 0$ for all u; then $\varphi(a') = a$, $\varphi(b') = b$.
(2) $\varphi'(u) < 0$ for all u; then $\varphi(a') = b$, $\varphi(b') = a$.

In the first case we say that φ preserves the orientation, and in the second case that φ changes the orientation.

Observe that if $\gamma: [a, b] \to U$ is a path piecewise of class C^1,

$$\gamma \circ \varphi: [a', b'] \to U$$

is also a path piecewise of class C^1. We say that the path $\gamma \circ \varphi$ is obtained from γ by the *change of parameter* φ.

PROPOSITION 3.3.1. With the preceding notation we have,

(3.3.1) $$\int_{\gamma\circ\varphi} \omega = \int_\gamma \omega \quad \text{if } \varphi \text{ preserves the orientation,}$$

(3.3.2) $$\int_{\gamma\circ\varphi} \omega = -\int_\gamma \omega \quad \text{if } \varphi \text{ changes the orientation.}$$

PROOF. It is sufficient to consider the case in which γ is of class C^1, the general case follows by taking a subdivision of $[a, b]$ which, by means of φ, corresponds to a subdivision of $[a', b']$. Suppose therefore that γ is of class C^1. By definition we have

$$\int_{\gamma\circ\varphi} \omega = \int_{a'}^{b'} (\gamma\circ\varphi)^*\omega.$$

Now (cf. Prop. 2.11.1) $(\gamma\circ\varphi)^*\omega = \varphi^*(\gamma^*\omega)$; if we put

$$\gamma^*\omega = f(t)\,dt,$$

then

$$\varphi^*(\gamma^*\omega) = f(\varphi(u))\varphi'(u)\,du,$$

giving

$$\int_{\gamma\circ\varphi} \omega = \int_{a'}^{b'} f(\varphi(u))\varphi'(u)\,du.$$

If φ preserves the orientation, then $\varphi(a') = a$, $\varphi(b') = b$, and the classical formula for the "change of variable in a simple integral" (see below) gives:

$$\int_{a'}^{b'} f(\varphi(u))\varphi'(u)\,du = \int_a^b f(t)\,dt,$$

which proves (3.3.1). If, on the contrary, φ changes the orientation, we have $\varphi(a') = b$, $\varphi(b') = a$, and the same formula gives

$$\int_{a'}^{b'} f(\varphi(u))\varphi'(u)\,du = \int_b^a f(t)\,dt = -\int_a^b f(t)\,dt,$$

which proves (3.3.2).

Remark. We shall recall below the proof of the formula for the change of variable in the case of a simple integral (cf. Remark following Corollary 3.4.2).

3.4 *The case where ω is the differential of a function*

Let $g\colon U \to F$ be a mapping of class C^1. Let the differential form ω be equal to the differential dg.

PROPOSITION 3.4.1. *If $\gamma\colon [a, b] \to U$ is a path piecewise of class C^1, then*

(3.4.1)
$$\boxed{\int_\gamma dg = g(\gamma(b)) - g(\gamma(a))}$$

(The right-hand side is equal to the difference in the values of g at the end and the beginning of the path γ.)

PROOF. It is sufficient to consider the case in which γ is of class C^1; the general case follows from this by subdivision of $[a, b]$. By definition, we have

$$\int_\gamma dg = \int_a^b \gamma^*(dg) = \int_a^b d(\gamma^*(g)) = \int_a^b d(g\circ\gamma).$$

Let $g \circ \gamma = h$; h is a function of class C^1: $[a, b] \to F$. Since $dh = h'(t)\,dt$, we finally obtain

$$\int_\gamma dg = \int_a^b h'(t)\,dt,$$

and the right-hand side is known to be equal to $h(b) - h(a)$. [Let us recall the proof of this result:

$$\int_a^\tau h'(t)\,dt$$

is a function of τ with derivative $h'(\tau)$, since the function h' is continuous; thus

$$\int_a^\tau h'(t)\,dt - h(\tau)$$

is constant (a function with zero derivative). Hence, from the fact that it takes the same values at $\tau = a$ and $\tau = b$, we obtain

$$\int_a^b h'(t)\,dt = h(b) - h(a).]$$

The proof of (3.4.1) is completed by replacing h by $g \circ \gamma$.

COROLLARY 3.4.2. If ω is the differential of a function g of class C^1, the integral $\int_\gamma \omega$ depends only on the origin $\gamma(a)$ and the extremity $\gamma(b)$ of the path γ.

Remark. Every continuous function f: $[a, b] \to F$ is equal to h', the derivative of the function h defined by

$$h(t) = \int_a^t f(u)\,du.$$

By a change of variable φ: $[a', b'] \to a, b$ of class C^1 such that $\varphi(a') = a$, $\varphi(b') = b$, we have

$$\int_a^b f(t)\,dt = h(b) - h(a) = h(\varphi(a')) - h(\varphi(b')) = \int_{a'}^{b'} (h \circ \varphi)'(u)\,du,$$

and since $(h \circ \varphi)'(u) = h'(\varphi(u) \circ \varphi'(u)) = f(\varphi(u)) \circ \varphi'(u)$, we obtain the "formula for the change of variable" which was used in § 3.3.

DEFINITION. Given a differential form ω: $U \to \mathscr{L}(E; F)$, every function f: $U \to F$ of class C^1, such that $df = \omega$, is called a *primitive* of ω.

In general a differential form of degree one *does not possess a primitive* (cf. § 2.12). If U is connected, the difference between two primitives of ω, f_1 and f_2, is constant, since $d(f_1 - f_2) = 0$.

Before stating the next theorem let us fix the terminology: a *loop* is a path whose origin and extremity coincide; a *polygonal loop* is a polygonal path whose origin and extremity coincide. (cf. *Differential Calculus*, Chap. 1, § 3.3).

THEOREM 3.4.3. *Let* U *be an open connected subset of a Banach space* E. *The following properties of a differential form* ω: $\mathscr{L}(E; F)$ *of class* C^0 *are equivalent:*

(a) ω possesses a primitive in U;

(b) $\int_\gamma \omega = 0$ for every loop γ, piecewise of class C^1, contained in U;

(c) $\int_\gamma \omega = 0$ for every polygonal loop γ contained in U.

Moreover, if U is starlike with respect to a point $x_0 \in U$, the preceding conditions are equivalent to

(d) $\int_\gamma \omega = 0$ for every triangle γ contained in U which has a vertex at x_0.

Note on (d). By "triangle" with vertex at x_0, we mean the perimeter of the triangle, i.e., a curve with origin and extremity at x_0 formed by three straight line segments. It is supposed that the sides of the triangle are contained in U, but not necessarily the portion of the plane within the triangle.

PROOF. (a) implies (b), by Prop. 3.4.1: for if $\gamma(b) = \gamma(a)$ then $\int_\gamma dg = 0$.

Clearly (b) implies (c) (for the set of polygonal loops form a subset of the set of loops). Let us show that (c) implies (a), so establishing the equivalence of (a), (b), (c). To do this, suppose that (c) is true; choose a point $x_0 \in U$; every $x \in U$ is the extremity of a polygonal path γ in U having origin at x_0, since U is connected (cf. *Differential Calculus*, Chap. 1, Prop. 3.3.5). The integral $\int_\gamma \omega$ does not depend on the choice of γ for a given x: since for two polygonal paths γ_1 and γ_2 with origin at x_0 and extremity at x, the difference

$$\int_{\gamma_1} \omega - \int_{\gamma_2} \omega$$

is equal to the integral of ω along the loop which is the union of γ_1 and γ_2, γ_2 being traversed in the inverse sense. This is zero, by (c). Let $f(x)$ be the common value of the integrals $\int_\gamma \omega$ taken along paths γ with origin x_0, extremity x. We shall show that the function f: U → F so defined has a derivative $f'(x)$ which is equal to ω, this will prove (a).

Let $[x, x + h]$ be the straight line segment from x to $x + h$; this is contained in U if $\|h\|$ is small enough, since U is open, and so contains a ball centre x. Denote by

$$\int_x^{x+h} \omega$$

the integral of ω along this straight line segment. Then

$$f(x + h) = f(x) + \int_x^{x+h} \omega, \tag{3.4.2}$$

since the path from x to $x + h$ is a polygonal path.

To calculate $\int_x^{x+h} \omega$ we make the change of variable

$$\gamma(t) = x + th \qquad (0 \leqslant t \leqslant 1):$$

by (3.2.2) we have:

$$f(x + h) - f(x) = \int_0^1 g(t)\, dt, \tag{3.4.3}$$

where

$$g(t) = \omega(x + th)\cdot h. \tag{3.4.4}$$

Since $\omega: U \to \mathscr{L}(E; F)$ is a continuous mapping, for every $\epsilon > 0$ there exists $\eta > 0$ such that

$$\|\omega(x + th) - \omega(x)\| \leqslant \epsilon \quad \text{for} \quad \|h\| \leqslant \eta$$

for every $t \in [0, 1]$. Hence

$$\|g(t) - g(0)\| \leqslant \epsilon\|h\| \quad \text{for} \quad \|h\| \leqslant \eta$$

for every $t \in [0, 1]$. Now

$$f(x + h) - f(x) - \omega(x)\cdot h = \int_0^1 (g(t) - g(0))\, dt,$$

so that

$$\|f(x + h) - f(x) - \omega(x)\cdot h\| \leqslant \epsilon\|h\| \quad \text{for} \quad \|h\| \leqslant \eta.$$

This proves that

$$\|f(x + h) - f(x) - \omega(x)\cdot h\| = o(\|h\|),$$

which shows that f is differentiable with derivative $f'(x)$ equal to $\omega(x)$. This establishes that (c) $\Rightarrow$ (a).

Suppose now that U is starlike with respect to x. It is evident that (c) implies (d), and if we can show that (d) implies (a) it will follow that (d) is equivalent to (a), (b), (c). Now, since U is starlike, we can *define*

$$f(x) = \int_{x_0}^{x} \omega, \tag{3.4.5}$$

the integral being taken along the straight line segment from x_0 to x. Now, if $\|h\|$ is small enough,

$$f(x + h) - f(x) = \int_x^{x+h} \omega,$$

the integral being taken along the straight line segment from x_0 to $x + h$: but this is exactly the hypothesis (d). As above, we conclude that f is differentiable with $f' = \omega$; thus (a) is true.

This completes the proof of Theorem 3.4.3.

Remark. When U is starlike with respect to the origin 0, the explicit form of (3.4.5) (with $x_0 = 0$) is

$$f(x) = \int_0^1 (\omega(tx)\cdot x)\, dt, \tag{3.4.6}$$

which is none other than Formula (2.13.1) used in the proof of Poincaré's theorem. This formula is therefore valid (in an open U, starlike with respect to the origin) *if the differential form ω is of class C^0 and possesses a primitive.* (In Poincaré's theorem we supposed that ω was of class C^1, and showed that if $d\omega = 0$, then ω possesses a primitive. Conversely, moreover, if ω is of class C^1 and possesses a primitive f, then f is of class C^2; therefore $d(df) = 0$, i.e., $d\omega = 0$.)

3.5 *The closed differential form of degree one*

DEFINITION. We say that a differential 1-form $\omega: U \to \mathscr{L}(E; F)$ (where U is an arbitrary open subset of a Banach space E) is *closed* if every point $x_0 \in U$ possesses an open neighbourhood V in which ω possesses a *primitive*. More briefly: we require that ω *possess a primitive locally.*

PROPOSITION 3.5.1. *In order that ω should be closed, it is necessary and sufficient that every point $x_0 \in U$ possess an open neighbourhood V, starlike with respect to x_0, such that*

$$\int_\gamma \omega = 0 \tag{3.5.1}$$

for every triangle γ contained in U. [Briefly we say: the integral (3.5.1) must be zero for every *sufficiently small* triangle γ contained in U.]

The proof follows directly from the definition and from (d) of Theorem 3.4.3.

THEOREM 3.5.2. *If ω is of class C^1, a necessary and sufficient condition for ω to be closed is that $d\omega = 0$.*

Indeed, we have just seen that, in an open starlike set, a necessary and sufficient condition for ω of class C^1 to possess a primitive is that $d\omega = 0$.

Let us recall that if U is an open subset of $\mathbf{R}^n$, ω may be written

$$\omega = \sum_{i=1}^{n} c_i(x)\, dx_i,$$

the coefficients c_i being functions of class C^1; the condition $d\omega = 0$ is then expressed by the relations

$$\frac{\partial c_i}{\partial x_j} = \frac{\partial c_j}{\partial x_i},$$

which are therefore necessary and sufficient for ω to be closed, i.e., to possess a primitive locally.

Note. A closed 1-form in an open $U \subset E$ (even if it is of class C^0) does not always possess a primitive throughout the whole of U (see, however, Theorem 3.8.1 below). Take, for example, $E = \mathbf{C}$, and let

$$U = \mathbf{C} - \{0\}$$

the complement of the origin. Let z denote the complex coordinate of a point of $\mathbf{C}$; this is a non-zero function in U. Let

$$\omega = \frac{1}{z}\, dz,$$

a differential form of degree 1, of class C^∞ in U. This is *closed*, for

$$d\omega = d\left(\frac{1}{z}\right) \wedge dz = -\frac{1}{z^2}\, dz \wedge dz = 0,$$

since $dz \wedge dz = 0$. To show that ω does not possess a primitive in U, it is sufficient to show that condition (a) of Theorem 3.4.3 is not satisfied. We shall exhibit a *loop* γ such that $\int_\gamma \omega \neq 0$; take the circle centre 0 of unit radius, given in parametric form by

$$z = e^{it}, \qquad 0 \leqslant t \leqslant 2\pi. \tag{3.5.2}$$

(We assume the reader to be familiar with the complex exponential.) The change of variable (3.5.2) transforms $\omega = (1/z)\,dz$ into

$$e^{-it}(ie^{it}\,dt) = i\,dt,$$

so that

$$\int_\gamma \frac{dz}{z} = \int_0^{2\pi} i\,dt = 2\pi i \neq 0.$$

Q.E.D.

If we set $z = x + iy$, then

$$\frac{dz}{z} = \frac{x\,dx + y\,dy}{x^2 + y^2} + i\frac{x\,dy - y\,dx}{x^2 + y^2},$$

hence

$$\int_\gamma \frac{x\,dy - y\,dx}{x^2 + y^2} = 2\pi.$$

3.6 *Primitive of a closed form along a path*

Let $\omega\colon \mathrm{U} \to \mathscr{L}(\mathrm{E}; \mathrm{F})$ be a *closed* differential 1-form of class C^0.

DEFINITION. For every path $\gamma\colon [a, b] \to \mathrm{U}$ (we assume that γ is *continuous*; it is unnecessary to assume that γ is piecewise of class C^1), we call a *primitive of ω along γ* any *continuous* function

$$f\colon [a, b] \to \mathrm{F}$$

satisfying the following condition: for every $t_0 \in [a, b]$, there exists an open neighbourhood V of $\gamma(t_0)$ in U, and a primitive F of ω in V, such that

$$\mathrm{F}(\gamma(t)) = f(t)$$

for all $t \in [a, b]$ close enough to t_0.

(*Remark.* In the definition we naturally suppose that $\mathrm{V} \subset \mathrm{U}$; further we can suppose that V is *connected*; every primitive of ω in V is then of the form F + constant.)

THEOREM 3.6.1. *If ω is a closed, differential 1-form in* U, *and γ is a continuous path in* U, *there exists a primitive f of ω along γ; such a primitive is uniquely defined to within an arbitrary additive constant.*

PROOF. First let us prove the uniqueness. Let f_1, f_2 be two primitives. If $t_0 \in [a, b]$, there exists a connected neighbourhood V of $\gamma(t_0)$ in which there are two primitives F_1 and F_2 of ω, such that

$$f_1(t) = \mathrm{F}_1(\gamma(t)), \qquad f_2(t) = \mathrm{F}_2(\gamma(t))$$

for t sufficiently close to t_0. Now $F_2 - F_1 =$ constant; therefore $f_2(t) - f_1(t)$ *is constant in a neighbourhood of* t_0. Thus the difference $f_2 - f_1$ is a continuous function on $[a, b]$ and *locally constant*. Now $[a, b]$ is a connected topological space. It follows that $f_2 - f_1$ is constant on $[a, b]$ (cf. *Differential Calculus*, Chap. 1, Lemma of § 3.3).

We now establish the *existence* of a primitive f of ω along γ. By definition, each point t_0 is contained in an open interval (open relative to $[a, b]$) within which there exists such a primitive. Since $[a, b]$ is compact, it can be covered by a finite number of such open intervals. Arrange these in an order such that if $I_1, \ldots, I_n$ denotes these intervals, then each I_k intersects the union of the preceding ones. By hypothesis, there exists a primitive f_k in I_k. In $I_2 \cap I_1$, $f_2 - f_1$ is constant, by the uniqueness theorem already proved. By adding a constant to f_2, we can arrange for $f_2 = f_1$ in $I_2 \cap I_1$; let g_2 be the function, in the interval $I_1 \cup I_2$, equal to f_1 in I_1, and f_2 in I_2; this is a primitive in $I_1 \cup I_2$. Next, in $(I_1 \cup I_2) \cap I_3$, $f_3 - g_2$ is constant; by adding a constant to f_3 we can arrange that $f_3 = g_2$ in $(I_1 \cup I_2) \cap I_3$, which defines a primitive g_3 in $I_1 \cup I_2 \cup I_3$. Proceeding in this way, we eventually obtain a primitive defined in the whole of $[a, b]$. This proves the theorem.

Remark 1. If the mapping γ is *constant*, it is evident that every primitive f along γ is *constant*.

Remark 2. Suppose that in the preceding theorem the path γ is of class C^1. Let

$$a = t_0 < t_1 < \cdots < t_n = b$$

be a subdivision such that, for each interval $[t_i, t_{i+1}]$, it is true that

$$f(t) = F_i(\gamma(t)) \quad \text{for} \quad t_i \leqslant t \leqslant t_{i+1},$$

where F_i is a primitive of ω in an open $V_i \subset U$ (such that $\gamma([t_i, t_{i+1}]) \subset V_i$). Then

$$\int_\gamma \omega = \sum_{i=0}^{n-1} \int \omega,$$

where γ_i denotes the restriction of γ: $[a, b] \to U$ to $[t_i, t_{i+1}]$. Now in V_i, $\omega = dF_i$, so that $\gamma^*(\omega) = d(\gamma^* F_i) = df$; hence

$$\int_{\gamma_i} \omega = \int_{t_i}^{t_{i+1}} df = f(t_{i+1}) - f(t_i).$$

Finally,

$$\int_\gamma \omega = \sum_{i=0}^{n-1} (f(t_{i+1}) - f(t_i)) = f(b) - f(a).$$

This result, true for a path γ of class C^1, is still true if γ is piecewise of class C^1: one first forms a subdivision, and then applies the result to each path of the subdivision. Thus:

PROPOSITION 3.6.2. *Let γ be a path piecewise of class* C^1, *ω a closed form, and f a primitive of ω along γ. Then*

$$\boxed{\int_\gamma \omega = f(b) - f(a)}. \tag{3.6.1}$$

In the general case where the path γ is only taken to be *continuous*, the *primitive* f exists and is unique to within an additive constant (Theorem 3.6.1). Thus $f(b) - f(a)$ is well defined. This leads us to *define* the integral $\int_\gamma \omega$ of a *closed* 1-form ω to be equal to $f(b) - f(a)$, the difference between the values at a and at b, of the primitive f of ω along γ. When the path γ is constant, the integral $\int_\gamma \omega = 0$. When γ is piecewise of class C^1, this new definition of $\int_\gamma \omega$ coincides with the earlier one, by Prop. 3.6.2. But the previous definition is also valid for a form ω which is not closed; on the other hand, *we are not concerned with the definition of* $\int_\gamma \omega$ *when the form* ω *is not closed and the path* γ *is only supposed continuous.*

3.7 *Homotopy of two paths*

For simplicity we shall only consider continuous paths described by a parameter which varies over $[0, 1]$ (the general case follows by change of parameter).

DEFINITION. We say that the two paths

$$\gamma_0: [0, 1] \to U, \qquad \gamma_1: [0, 1] \to U$$

are *homotopic* if there exists a *continuous* mapping δ of the square

$$0 \leqslant t \leqslant 1, \qquad 0 \leqslant u \leqslant 1$$

(a compact subset of $\mathbf{R}^2$ with coordinates t and u), with values in U, such that

$$\delta(t, 0) = \gamma_0(t), \qquad \delta(t, 1) = \gamma_1(t) \quad \text{for} \quad 0 \leqslant t \leqslant 1. \tag{3.7.1}$$

The mapping δ is called the *homotopy* of γ_0 to γ_1. If this is the case, each $u \in [0, 1]$ defines a path

$$\gamma_u: [0, 1] \to U$$

given by

$$\gamma_u(t) = \delta(t, u).$$

We sometimes say that the family of paths γ_u defines a *continuous deformation* of the path γ_0 into the path γ_1.

The relation "γ_0 is homotopic to γ_1" is an equivalence relation in the set of paths in U. (*Exercise:* prove this.)

Homotopy when the origin and extremity are fixed: when γ_0 and γ_1 have the same origin $\gamma_0(0) = \gamma_1(0)$, and the same extremity $\gamma_0(1) = \gamma_1(1)$, we say that δ *is an homotopy with origin and extremity fixed* if, apart from (3.7.1) δ also satisfies

$$\delta(0, u) = \gamma_0(0), \qquad \delta(1, u) = \gamma_0(1) \tag{3.7.2}$$

for every $u \in [0, 1]$; in other words, we require that, for every u, the path γ_u has the same origin as γ_0 and γ_1, and the same extremity as γ_0 and γ_1.

THEOREM 3.7.1. *Let ω be a closed differential 1-form in an open* $U \subset E$; *let*

$$\gamma_0: [0, 1] \to U, \qquad \gamma_1: [0, 1] \to U$$

be two continuous paths with the same origin and the same extremity. If γ_0 *and* γ_1 *are homotopic with the origin and extremity fixed, then*

$$\int_{\gamma_0} \omega = \int_{\gamma_1} \omega \tag{3.7.3}$$

(where the "integrals" are defined as in § 3.6).

This theorem follows from a second theorem which concerns the existence of a "primitive of ω following a continuous mapping of a rectangle" (cf. definition below).

DEFINITION. Let δ be a continuous mapping of the rectangle (R)

$$a \leqslant t \leqslant b, \qquad a' \leqslant u \leqslant b'$$

with values in an open $U \subset E$, and let ω be a *closed* differential 1-form in U. A *continuous* mapping $f\colon \delta \to F$ is called a *primitive of* ω *following* δ if, for every point (t_0, u_0) of the rectangle (R), there exists an open connected neighbourhood V of $\delta(t_0, u_0)$ contained in U, and a primitive F of ω in V, such that

$$F(\delta(t, u)) = f(t, u)$$

for every point $(t, u) \in (R)$ sufficiently close to (t_0, u_0).

The reader will observe that this definition is based on the corresponding one for a "primitive of ω along a path".

THEOREM 3.7.2 (analogue of Theorem 3.6.1). *With the preceding notation, there exists a primitive f of* ω *following the mapping* δ*; such a primitive is uniquely defined to within an arbitrary additive constant.*

This theorem will be proved later. First we shall see how it implies Theorem 3.7.1. Let γ_0 and γ_1 be two paths in U with the same origin and same extremity, and suppose that γ_0 and γ_1 are homotopic with the origin and extremity fixed. There exists a continuous mapping δ of the square

$$0 \leqslant t \leqslant 1, \qquad 0 \leqslant u \leqslant 1$$

contained in U, satisfying (3.7.1) and (3.7.2). By Theorem (3.7.2) there exists a primitive f of ω following δ. The mappings

$$f_0(t) = f(t, 0), \qquad f_1(t) = f(t, 1)$$

are evidently primitives of ω along γ_0 and γ_1 respectively (see the definition of a primitive along a path). The relation (3.7.3) to be established thus takes the form

$$f(1, 0) - f(0, 0) = f(1, 1) - f(0, 1),$$

i.e.,

$$f(0, 1) - f(0, 0) = f(1, 1) - f(1, 0). \tag{3.7.4}$$

We shall prove this by showing that each side of (3.7.4) is zero: the function $u \mapsto f(0, u)$

is a primitive of ω along the path $u \mapsto \delta(0, u)$; this path is *constant* by (3.7.2), therefore the primitive is constant, so that

$$f(0, 1) - f(0, 0) = 0.$$

Similarly it may be shown that $f(1, 1) - f(1, 0) = 0$.

Hence we have shown that Theorem 3.7.1 is a consequence of Theorem 3.7.2.

PROOF OF THEOREM 3.7.2. This is analogous to the proof of Theorem 3.6.1, although a little more complicated. We show first that if f_1 and f_2 are two primitives of ω following δ, the difference $f_2 - f_1$ is *constant* in the rectangle (R); to do this, observe as in Theorem 3.6.1, that $f_2 - f_1$ is *locally constant*, and then, that the rectangle (R) is *connected* (it is the "product" of two connected intervals).

It remains to prove the *existence* of a primitive f of ω following δ. We already know (from the definition of a primitive) that each point $(t_0, u_0) \in$ (R) possesses an open neighbourhood in which such a primitive exists. Since (R) is compact, there exists a sufficiently fine "net" of (R), defined by the two subdivisions

$$a = t_0 < t_1 < \cdots < t_n = b$$

$$a' = u_0 < u_1 < \cdots < u_p = b',$$

such that every mesh

$$t_i \leqslant t \leqslant t_{i+1}, \qquad u_j \leqslant u \leqslant u_{j+1} \qquad (0 \leqslant i < n, 0 \leqslant j < p)$$

possesses a neighbourhood in which there exists a primitive f of ω following the mapping δ (restricted to this neighbourhood). Therefore there exists $\epsilon > 0$ (independent of i, j, which take only a finite number of values) such that there exists a primitive f_{ij} in that part of the rectangle (R) defined by the inequalities

$$t_i - \epsilon < t < t_{i+1} + \epsilon, \qquad u_j - \epsilon < u < u_{j+1} + \epsilon.$$

Take first $j = 0$, and allow i to vary from 0 to $n - 1$; in a stepwise manner, we can add constants to each of the functions $f_{0,0}, f_{1,0}, \ldots, f_{n-1,0}$ so that together they define a primitive g_0 in the rectangle

$$a \leqslant t \leqslant b \qquad a' \leqslant u \leqslant u_1 + \epsilon.$$

In the same way we obtain a primitive g_1 in the rectangle

$$a \leqslant t \leqslant b, \qquad u_1 - \epsilon < u \leqslant u_2 + \epsilon,$$

etc. Finally we obtain a sequence of primitives $g_0, g_1, \ldots, g_{p-1}$; by adding suitable constants to each of these, they may be made to define a primitive f in the whole of the rectangle (R).

With the aid of Theorem 3.7.2 we can now establish the analogue of Theorem 3.7.1. First let us give the

DEFINITION. Let γ_0 and γ_1 be two *loops* contained in U (so that $\gamma_0(0) = \gamma_0(1)$, $\gamma_1(0) =$

$\gamma_1(1)$); we say that they are *homotopic* if there exists a continuous mapping δ of the square

$$0 \leqslant t \leqslant 1, \qquad 0 \leqslant u \leqslant 1$$

contained in U, which, apart from (3.7.1), also satisfies

(3.7.5) $$\delta(0, u) = \delta(1, u) \quad \text{for all} \quad u \in [1, 0].$$

The latter condition shows that, for each u, the path $\gamma_u(t) = \delta(t, u)$ is a *loop*, i.e., its extremity coincides with its origin.

Suppose that ω is a closed differential 1-form in U. Let us apply Theorem 3.7.2 to the mapping δ of the definition; let f be a primitive of ω following δ. The two paths

$$u \mapsto \delta(0, u) \quad \text{and} \quad u \mapsto \delta(1, u)$$

are equal, so that

$$f(0, 1) - f(0, 0) = f(1, 1) - f(1, 0),$$

therefore equation (3.7.4) is valid in this case. As before, this implies that

$$\int_{\gamma_0} \omega = \int_{\gamma_1} \omega.$$

We have thus established a variant of Theorem 3.7.1:

THEOREM 3.7.3. *Let ω be a closed differential 1-form in an open* $U \subset E$; *if two loops γ_0 and γ_1 contained in* U *are homotopic, then*

$$\int_{\gamma_0} \omega = \int_{\gamma_1} \omega.$$

Note. The relation of homotopy between two loops of U is an equivalence relation. If ω is a closed form in U, and γ a loop of U, we have seen (end of § 3.4) that $\int_\gamma \omega$ is not necessarily zero. But Theorem 3.7.3 states that the value of $\int_\gamma \omega$ depends only on the *class of homotopy of the loop* γ. In particular, if a loop is homotopic to a *point*, then $\int_\gamma \omega = 0$ for every closed 1-form ω. (A point is defined by a constant mapping.)

3.8 *Simply-connected domains*

DEFINITION. A topological space X is said to be *simply connected* if: (1) for every pair of points $(x_0, x_1) \in X$ there exists a continuous path $\gamma\colon [0, 1] \to X$ joining x_0 to x_1; (2) every loop in X is homotopic to a point.

It is clear that if two topological spaces X and Y are homeomorphic and if one of them is simply-connected, then the other is also.

If we apply this definition to an open subset U of a Banach space E, condition 1 signifies that U is connected (*Differential Calculus*, Chap. 1, Prop. 3.3.5). If, further, U satisfies condition 2, then $\int_\gamma \omega = 0$ for every loop γ in U and every closed 1-form ω in U (by Theorem 3.7.3). By using Theorem 3.4.3 (the equivalence of (a) and (b)), we obtain:

THEOREM 3.8.1. *Let* $\omega: U \to \mathscr{L}(e; F)$ *be a closed differential* 1*-form. If* U *is simply-connected, then* ω *possesses a primitive in* U.

Examples of simply-connected domains (a domain is an open set): every open, starlike U is simply-connected. Indeed, suppose for a firm basis, that U is starlike with respect to the origin 0; if x_0 and x_1 are two points of U, the polygonal path formed by the segment $[x_0, 0]$ and the segment $[0, x_1]$ begins at x_0 and ends at x_1, thus U satisfies condition 1 (i.e., it is connected). Further, if $\gamma: [0, 1] \to U$ is a closed curve; put $\delta(t, u) = u \cdot \gamma(t)$ (the product of $\gamma(t) \in E$ with the scalar $u \in [0, 1]$); $\delta(t, u) \in U$ since U is starlike with respect to 0; the mapping δ is continuous, and shows that the loop

$$t \mapsto \delta(t, 1) = \gamma(t)$$

is homotopic to the point

$$t \mapsto \delta(t, 0) = 0.$$

In particular, every *connected* domain is simply-connected (since it is starlike with respect to each of its points).

By a remark already made, every open $U \subset E$ which is homeomorphic to an open starlike set is simply-connected.

On the other hand, take $E = \mathbf{C}$, and $U = \mathbf{C} - \{0\}$; U is not *simply-connected* (although it *is* connected), since we have seen (end of § 3.5) that the closed differential form $(1/z)\, dz$ does not possess a primitive in U.

Exercise. Show that for a topological space X satisfying condition 1, the following four properties are equivalent:

(a) X is simply-connected.

(b) Every continuous mapping of the circle $|z| = 1$ (z denoting a complex number in $\mathbf{C}$) with values in X can be extended to a continuous mapping of the disc $|z| \leqslant 1$ into X.

(c) Every continuous mapping of the sides of a square with values in X can be extended to a continuous mapping of the square into X.

(d) If two paths of X have the same origin and the same end point, they are "homotopic with the origin and extremity fixed" (cf. § 3.7).

4. Integration of differential forms of degree >1

4.1 *Differentiable partitions of unity*

We propose to prove the following theorem:

THEOREM 4.1.1. *Let* K *be a compact subset of* $\mathbf{R}^n$, *and let* $(U_i)_{i \in I}$ *be a finite covering of* K *by the open sets* $U_i \subset \mathbf{R}^n$. *Then there exist functions*

$$f_i: \mathbf{R}^n \to [0, 1] \qquad (i \in I)$$

of class C^∞, *such that:*

(i) $\operatorname{supp} f_i \subset U_i$

(supp f_i denotes the "support" of the function f_i, i.e., the closed set of $x \in \mathbf{R}^n$ such that

$f_i(x) \neq 0$; the complement of supp f_i is the largest open set in which f_i is identically zero).

(ii) $\sum_{i \in I} f_i(x) \leqslant 1$ for all $x \in \mathbf{R}^n$.

(iii) $\sum_{i \in I} f_i(x) = 1$ for all $x \in K$.

Note. Conditions (ii) and (iii) are expressed by saying that the f_i define a "differentiable partition of unity" over K; condition (i) says that the partition f_i is *subordinate* to the open covering (U_i).

Let us indicate an important particular case of Theorem 4.1.1: that in which the set I contains only one element:

COROLLARY 4.1.2. *Let* K *be a compact subset of* $\mathbf{R}^n$, *and let* U *be an open set containing* K. *There exists a function*

$$f: \mathbf{R}^n \to [0, 1]$$

of class C^∞, *such that*

$$\operatorname{supp} f \subset U, \qquad f(x) = 1 \quad \text{for} \quad x \in K.$$

In the following proof of Theorem 4.1.1, an essential rôle is played by the function (defined in $\mathbf{R}^n$, and taking values in [0, 1]), defined by:

$$\lambda(x) = \begin{cases} \exp\left(\dfrac{1}{\|x\|^2 - 1}\right) & \text{for} \quad \|x\| < 1 \\ 0 \quad \text{for} \quad \|x\| \geqslant 1, \end{cases} \tag{4.1.1}$$

where $\|x\|$ denotes the Euclidean norm in $\mathbf{R}^n$. We show that this function λ is of class C^∞: this is obvious for $\|x\| > 1$; it is also evident for $\|x\| < 1$, for $\|x\|^2$ is a function of class C^∞ (sum of the squares of the coordinates of x). Thus we have to show that if $a \in \mathbf{R}^n$ and $\|a\| = 1$, then the function $\lambda(x)$ is infinitely differentiable at the point a.

By induction on the integer $k \geqslant 0$, we shall show that the kth derivative at the point a is zero; this is trivial for $k = 0$, since by definition $\lambda(a) = 0$ for $\|a\| = 1$; suppose it is true for $k - 1$ $(k \geqslant 1)$; if $\lambda^{(k)}(a)$ exists and is zero, then

$$\|\lambda^{(k-1)}(x)\| = o(\|x - a\|) \tag{4.1.2}$$

when x tends to a (since $\lambda^{(k-1)}(a) = 0$). Now (4.1.2) is obviously true for $\|x\| \geqslant 1$, since in this case $\lambda^{(k-1)}(x) = 0$; in fact it is true for $\|x\| > 1$ since, by (4.1.1), the function $\lambda(x)$ is identically zero for $\|x\| > 1$; and it is true for $x = 1$ by the induction hypothesis. Thus it remains to prove (4.1.2) when x tends to a for $\|x\| \leqslant 1$. In this case one can, at least theoretically, calculate successively the derivatives of the function λ given by the upper line in the definition (4.1.1). By induction on k, the reader may prove, as an exercise, that for $\|x\| < 1$,

$$\lambda^{(k-1)}(x) \in \mathscr{L}_{k-1}(\mathbf{R}^n; \mathbf{R})$$

has a norm

$$\|\lambda^{(k-1)}(x)\| \leqslant \frac{m_k}{(1 - \|x\|^2)} \exp\left(\frac{1}{\|x\|^2 - 1}\right),$$

where m_k is a number independent of x; hence

$$\|\lambda^{(k-1)}(x)\| = o(1 - \|x\|^2).$$

Now $1 - \|x\|^2 = \|a\|^2 - \|x\|^2 \leqslant 2(\|a\| - \|x\|) \leqslant 2\|x - a\|$, which finally gives (4.1.2).

Thus the definition (4.1.1) gives a well-defined infinitely differentiable function for all $x \in \mathbf{R}^n$. At each point x of the open disc $\|x\| < 1$, one has $\lambda(x) > 0$; therefore in every open disc $\|x\| < r (r < 1)$, $\lambda(x)$ has a lower bound $m(r) > 0$; indeed,

(4.1.3)
$$m(r) = \exp\left(\frac{1}{r^2 - 1}\right).$$

Now choose a point $x_0 \in \mathbf{R}^n$ and a number $r > 0$; the function

$$x \mapsto \lambda\left(\frac{x - x_0}{r}\right)$$

is zero outside the ball $\|x - x_0\| \leqslant r$, and strictly positive inside this ball. It is of class C^∞.

Thus prepared, we shall now prove two lemmas which will lead us to the proof of Theorem 4.1.1.

Lemma 1. There exists a function

$$\mu: \mathbf{R}^+ \to [0, 1]$$

of class C^∞ which possesses the following properties:

$$\mu(0) = 0, \qquad \mu(t) = 1 \quad \text{for} \quad t \geqslant 1.$$

($\mathbf{R}^+$ the semi-axis $t \geqslant 0$).

PROOF. Take

(4.1.4)
$$\mu(t) = \begin{cases} 1 - \exp(t^2/(t^2 - 1)) & \text{for} \quad t < 1 \\ 1 \quad \text{for} \quad t \geqslant 1. \end{cases}$$

This function is indeed of class C^∞, since

$$\mu(t) = 1 - e \cdot \lambda(t)$$

Lemma 2. As in the statement of Theorem 4.1.1, given a compact K and a finite covering $(U_i)_{i \in I}$ of K, there exist functions

$$g_i: \mathbf{R}^n \to \mathbf{R}^+$$

of class C^∞ such that

(a) supp $g_i \subset U$;

(b) $\sum_{i \in I} g_i(x) \geqslant 1$ for every $x \in K$.

PROOF OF LEMMA 2. To each $x \in K$ associate an index $i = i(x) \in I$ such that $x \in U_i$. Choose $r(x) > 0$ such that the closed ball of centre x and radius $r(x)$ is contained in U_i; since, when x varies over K, the open balls $B(x, \frac{1}{2}r(x))$ cover the *compact* K, we can cover

K with a finite number of such balls. Let x_α be the centre of a ball (where α varies over a finite set A), and set $r(x_\alpha) = r_\alpha$. Thus:

1. The open balls $B(x_\alpha, \frac{1}{2}r_\alpha)$ cover K.
2. Each closed ball centre x_α and of radius r_α is contained in the open $U_{i(x_\alpha)}$.

Let λ_α be the function

$$x \mapsto \frac{1}{m(\frac{1}{2})}\lambda\left(\frac{x - x_\alpha}{r_\alpha}\right);$$

it is of class C^∞, and takes values in $\mathbf{R}^+$; its support is contained in $U_{i(x_\alpha)}$, and it is $\geqslant 1$ in the open ball $B(x_\alpha, \frac{1}{2}r_\alpha)$. Thus for every point $x \in K$ there exists at least one index α such that $\lambda_\alpha(x) \geqslant 1$; *a fortiori* the sum

$$\sum_{\alpha \in A} \lambda_\alpha(x)$$

is a function which is $\geqslant 1$ at every point of K. For each $i \in I$, let A_i be the set of $\alpha \in A$ such that $i(x_\alpha) = i$; the A_i, when i varies over I, forms a *partition* of A. Put, for $x \in \mathbf{R}^n$

(4.1.5) $$g_i(x) = \sum_{\alpha \in A_i} \lambda_\alpha(x).$$

The support of g_i is evidently contained in the union of the supports of the functions λ_α for $\alpha \in A_i$, and as each of these supports is contained in U_i, it follows that

$$\operatorname{supp} g_i \subset U_i.$$

Further, $\sum_{i \in I} g_i(x) = \sum_{\alpha \in A} \lambda_\alpha(x) \geqslant 1$, at every point $x \in K$. This proves Lemma 2.

We are now able to prove Theorem 4.1.1. Let us begin by proving the Corollary 4.1.2: given K and U as in the statement of the corollary, we apply Lemma 2 (taking for I a set of one element), to obtain a function

$$g: \mathbf{R}^n \to \mathbf{R}^+$$

of class C^∞, such that

$$\operatorname{supp} g \subset U, \qquad g(x) \geqslant 1 \quad \text{for} \quad x \in K.$$

Now use the function μ of Lemma 1, and put

$$f = \mu \circ g: \mathbf{R}^n \to [0, 1].$$

It is evident that f satisfies the conditions in the statement of Corollary 4.1.2, which is thus proved.

Now consider the *general case*, Theorem 4.1.1. Associate with the open covering $(U_i)_{i \in I}$ the functions g_i of Lemma 2, and put

$$g(x) = \sum_{i \in I} g_i(x).$$

The set U of $x \in \mathbf{R}^n$ such that $g(x) > 0$ contains K and is *open* (since g is continuous). By Corollary 4.1.2 (already proved), there exists

$$f: \mathbf{R}^n \to [0, 1]$$

of class C^∞, such that

$$\operatorname{supp} f \subset U, \qquad f(x) = 1 \quad \text{for} \quad x \in K.$$

For each $i \in I$ put:

$$f_i(x) = \begin{cases} \dfrac{g_i(x)}{g(x)} f(x) & \text{for } x \in U \\ 0 & \text{for } x \notin U. \end{cases} \tag{4.1.6}$$

Since $g(x) \neq 0$ for $x \in U$, $f_i(x)$ is well defined. It remains to show that the functions f_i are of class C^∞ and satisfy conditions (i), (ii) and (iii) of Theorem 4.1.1. Now in the neighbourhood of the point $x_0 \in U$, f_i is the quotient of two functions of class C^∞ (the denominator being >0), therefore f_i is of class C^∞; in the neighbourhood of a point $x_0 \notin U$, f_i is identically zero, since $x \notin \operatorname{supp} f_i$, thus f_i is identically zero in a neighbourhood of x_0, and so $f_i(x)$ is identically zero.

Thus f_i is of class C^∞ in the neighbourhood of every point $x_0 \in \mathbf{R}^n$. The value $f_i(x)$ is evidently $\geqslant 0$ at every point $x \in \mathbf{R}^n$. Property (i) is true, since $\operatorname{supp} f_i \subset \operatorname{supp} g_i$. Property (ii) is evident, since for $x \in U$

$$\sum_{i \in I} f_i(x) = f(x),$$

since $\sum_{i \in I} g_i(x) = g(x)$. Finally, if $x \in K$, $f(x) = 1$, which proves (iii).

This completes the proof of Theorem 4.1.1.

4.2 *Compact sets with boundary in the plane* $\mathbf{R}^2$

Let x and y denote the coordinates in $\mathbf{R}^2$.

DEFINITION. We say that a *compact* subset $L \subset \mathbf{R}^2$ is a *curve of class* C^1 if each point $(x_0, y_0) \in L$ possesses an open neighbourhood U such that the set $L \cap U$ of points situated in L and in U can be defined by an equation of the form

$$y = \varphi(x)$$

(where φ is a function of class C^1 in a neighbourhood of x_0, such that $\varphi(x_0) = y_0$), *or* by an equation of the form

$$x = \psi(y)$$

(where ψ is a function of class C^1 in a neighbourhood of y_0, such that $\psi(y_0) = x_0$).

Exercise. Show that it comes to the same thing to say that there exists *a path of class* C^1 (cf. § 3.1)

$$\gamma\colon [a, b] \to \mathbf{R}^2$$

such that the derivative $\gamma'(t)$ is $\neq 0$ for every $t \in [a, b]$, and that (x_0, y_0) is the image $\gamma(t_0)$ of a $t_0 \in]a, b[$, and that there exists an open $U \ni (x_0, y_0)$ such that the set of points of the image of γ contained in U coincides with $L \cap U$.

The equivalence of the two definitions is a simple consequence of the implicit function theorem. Note that the condition $\gamma'(t)_0 \neq 0$ implies that the mapping γ is injective for t sufficiently close to t_0.

PROPOSITION 4.2.1. *Let* $L \subset \mathbf{R}^2$ *be a curve of class* C^1. *In the neighbourhood of each of its points,* L *divides the plane into two regions.*

This means that we can choose an open $V \subset U$, which contains (x_0, y_0), is connected, and such that $V - V \cap L$ has two components, each of which is connected. This may be seen, for example, by supposing that L is defined in the neighbourhood of (x_0, y_0) by an equation $y = \varphi(x)$. Choose $\epsilon > 0$, and $\eta > 0$ such that:

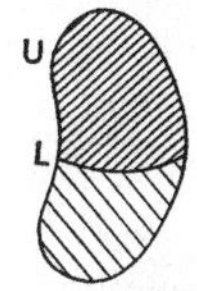

1. The open rectangle $|x - x_0| < \epsilon$, $|y - y| < \eta$ is contained in U;
2. $|\varphi(x) - y_0| < \eta$ for every x such that $|x - x_0| < \epsilon$.

If V denotes this open rectangle, V is connected; $V - V \cap L$ has two components each of which is connected, viz., the set of points $(x, y) \in V$ such that $y < \varphi(x)$, and the set of points $(x, y) \in V$ such that $y > \varphi(x)$.

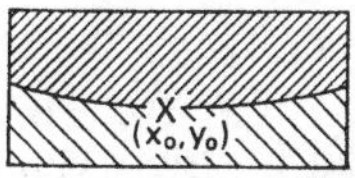

DEFINITION. We say that the compact set $L \subset \mathbf{R}^2$ is a *curve piecewise of class* C^1 if each point (x_0, y_0) of L possesses an open neighbourhood U such that the set $L \cap U$ coincides with the set of points of U which belong to the origin of a path

$$\gamma: [a, b] \to \mathbf{R}^2$$

which is piecewise of class C^1 (cf. § 3.1) and has the following properties: the mapping γ is injective, and in each of the partial (closed) intervals where γ is of class C^1, its derivative $\gamma'(t)$ is $\neq 0$.

The points (x_0, y_0) of L in the neighbourhood of which L is not of class C^1 are called the *angular points* of L; they are *isolated*, and therefore of finite number, since L is compact. The remaining points of L are called the *regular* points of L. To say that L is a curve of class C^1 is to say that all the points of L are regular.

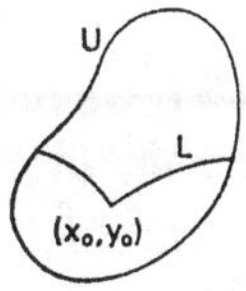

DEFINITION. We say that the *compact* subset $K \subset \mathbf{R}^2$ is a *compact set with boundary* if it possesses the following properties:

(a) The set ∂K of frontier points of K is a curve piecewise of class C^1.

(b) For every *regular* point $(x_0, y_0) \in \partial K$ there exists an open connected neighbourhood V of (x_0, y_0) such that $V - V \cap (\partial K)$ has *two* components each of which is connected, one of these being formed from points in the *interior* of K, and the other of points complementary to K.

Briefly, condition (a) says that in the neighbourhood of every frontier point of K, the points of K are situated on one side of the boundary ∂K, and the points of the complement $\complement K$ are situated on the other side of ∂K.

Remark. The boundary ∂K is not necessarily connected, even if K is connected.

Examples of boundary compact sets:

Orientation of the boundary of K. In the neighbourhood of every regular point of the boundary ∂K, ∂K coincides with the image of a path of class C^1. Now a path can be oriented in two different manners. Each parametrization defines an orientation; two parametrizations define the same orientation if the change of parameter is defined by a strictly increasing function, and define opposite orientations if the change of parameter is defined by a strictly decreasing function. Now the condition (b) says that in the neighbourhood of each regular point of ∂K, K is situated on one side of ∂K. We choose the orientation of ∂K (i.e., the sense in which it is traversed) in such a way that K is on the *left* as ∂K is traversed. More precisely, the tangent vector $\gamma'(t_0)$ to the curve ∂K, parametrized by t in the neighbourhood of t_0 (with $\gamma(t_0) = (x_0, y_0)$) must be in such a direction that every vector making an angle $+\pi/2$ with $\gamma'(t_0)$ is directed into the interior of K. Let us explain this explicitly when, in the neighbourhood of (x_0, y_0), ∂K is defined by

$$y = \varphi(x);$$

then, if the interior of K, in the neighbourhood of (x_0, y_0), is defined by $y > \varphi(x)$, we take on ∂K (in the neighbourhood of (x_0, y_0)) the orientation defined by increasing x (i.e., we take x as the parameter). If, on the contrary, the interior of K satisfies $y < \varphi(x)$, we take $-x$ as the parameter, so that the orientation is in the direction of x decreasing.

If we proceed thus at all the regular points of the boundary ∂K, it is clear that ∂K is the union of a finite number of *oriented paths*. We can therefore define the curvilinear integral $\int_{\partial K} \omega$ for a differential form ω of degree one. When the orientation is so defined, ∂K is called the *oriented boundary* of the compact K.

4.3 *Integral of a differential 2-form on a compact set with boundary*

Let ω be a differential form of degree 2, of class C^0, in an open $U \supset K$, and taking values in a Banach space F:

$$\omega \in \Omega_2^{(0)}(U, F).$$

Remark. In the neighbourhood of K, ω coincides with a differential form defined in the whole of the $\mathbf{R}^2$ plane: in fact, by Corollary 4.1.2, there exists a function $\lambda: \mathbf{R}^2 \to [0, 1]$, of class C^∞ (and *a fortiori* continuous), equal to 1 on a compact neighbourhood of K, and whose support is contained in U. We then take for α the differential form equal to $\lambda\omega$ in U, and to 0 in the complement of the support of λ.

Using, as always, x, y as coordinates in $\mathbf{R}^2$, the differential form ω has the canonical form

$$\omega = f(x, y)\, dx \wedge dy,$$

where f is a continuous mapping $U \to F$. Let $\bar{f}(x, y)$ be the function $\mathbf{R}^2 \to F$ defined as follows:

$$\begin{cases} \bar{f}(x, y) = f(x, y) & \text{for} \quad (x, y) \in K \\ \bar{f}(x, y) = 0 & \text{for} \quad (x, y) \notin K. \end{cases} \tag{4.3.1}$$

The function $\bar{f}$ has a compact support and is Lebesgue-integrable; in fact the characteristic function χ_K of every compact K is Lebesgue-integrable, therefore the product $\chi_K f = \bar{f}$ is Lebesgue-integrable. Indeed it can be shown that χ_K is Riemann-integrable since K is a compact set with boundary; therefore $\bar{f}$ is Riemann-integrable.

The integral

$$\iint \bar{f}(x, y)\, dx \wedge dy$$

with respect to the element of area $dx \wedge dy$ of the plane is therefore defined. By definition it is denoted by

$$\iint_K f(x, y)\, dx \wedge dy;$$

and also by definition, this is the integral $\iint_K \omega$. The value of this integral is an element of F.

We shall give, without proof, several properties of the preceding integral:

(i) For given K, the mapping $\omega \mapsto \iint_K \omega$ is a *linear* function of $\omega \in \Omega_2^{(0)}(U, F)$.

(ii) If K′ is a compact set contained in K, and such that

$$(\text{supp } \omega) \cap K \subset K', \tag{4.3.2}$$

then

$$\iint_K \omega = \iint_{K'} \omega \tag{4.3.3}$$

(supp ω denotes the *support* of $\omega = f\, dx \wedge dy$, i.e., the support of the function f, which is *closed* in U; its intersection with K is compact. It is evident that (4.3.2) implies (4.3.3), for the function $\bar{f}$ defined by (4.3.1) is equal to f on K′, and equal to 0 outside K′).

(iii) The integral

$$\iint \bar{f}(x, y)\, dx \wedge dy$$

can be calculated by two successive integrations: for y fixed, one first calculates

$$\int \bar{f}(x, y)\, dx,$$

which exists in the sense of Lebesgue; it is a function $g(y)$, which is Lebesgue-integrable, and such that $\int g(y)\, dy$ is equal to $\iint \bar{f}(x, y)\, dx \wedge dy$:
i.e.,

$$\iint \bar{f}(x, y)\, dx \wedge dy = \int dy\left(\int \bar{f}(x, y)\, dx\right). \tag{4.3.4}$$

Similarly,

$$\iint \bar{f}(x, y)\, dx \wedge dy = \int dx\left(\int \bar{f}(x, y)\, dy\right) \tag{4.3.5}$$

(iv) If

$$\begin{cases} x = x_0 + u\cos\theta - v\sin\theta = a(u, v) \\ y = y_0 + u\sin\theta + v\cos\theta = b(u, v) \end{cases}$$

is a *direct Euclidean* displacement (composed of a rotation and a translation), then

$$\iint \bar{f}(x, y)\, dx \wedge dy = \iint \bar{f}(a(u, v), b(u, v))\, du \wedge dv. \tag{4.3.6}$$

If φ is the mapping

$$(u, v) \mapsto (a(u, v), b(u, v))$$

of $\mathbf{R}^2$ into $\mathbf{R}^2$, and K' is the compact $\varphi^{-1}(K)$, then (4.3.6) may also be written in the form

$$\iint_K \omega = \iint_{K'} \varphi^*(\omega), \tag{4.3.7}$$

which is only a particular case of the formula for the "change of variable" which will be established later. Note that we do indeed have

$$\varphi^*(dx \wedge dy) = du \wedge dv,$$

since the Jacobian $\partial(x, y)/\partial(u, v) = \sin^2\theta + \cos^2\theta = 1$ (the determinant of a rotation).

The proof of property (iv) follows immediately if one admits the invariance of an area element under direct displacement.

4.4 *Stokes's theorem in the plane*

THEOREM 4.4.1. *Suppose* $K \subset \mathbf{R}^2$ *is a compact set with boundary and let* α *be a differential* 1*-form of class* C^1 *in an open* $U \supset K$, *with values in* F (*a Banach space*). *Then*

$$\boxed{\iint_K d\alpha = \int_{\partial K} \alpha}, \tag{4.4.1}$$

the curvilinear integral $\int_{\partial K} \alpha$ *being taken along the oriented boundary of the compact* K (with the convention of orientation defined in § 4.2).

More explicitly, by writing in canonical form

$$\alpha = P(x, y)\, dx + Q(x, y)\, dy,$$

where P and Q are functions of class C^1, defined in U with values in F, (4.4.1) becomes, noting that

$$d\alpha = \left(\frac{\partial Q}{\partial x} - \frac{\partial P}{\partial y}\right) dx \wedge dy,$$

$$\boxed{\iint_K \left(\frac{\partial Q}{\partial x} - \frac{\partial P}{\partial y}\right) dx \wedge dy = \int_{\partial K} P\, dx + Q\, dy}. \tag{4.4.2}$$

This form is known as the *Green–Riemann formula*.

The proof will occupy several stages, and will be given in § 4.5. Meanwhile we shall define an auxiliary notion:

DEFINITION. A rectangle† R in the plane $\mathbf{R}^2$ is said to be K-*privileged* if it satisfies one of the following conditions:

(a) R does not meet the boundary ∂K.

(b) The intersection $R \cap (\partial K)$ is a curve of class C^1 joining two opposite corners of R such that there exists a rotation transforming R into a rectangle whose sides are parallel to the axes, and $R \cap \partial K$ into a curve defined by

$$y = \varphi(x), \quad \text{and also by} \quad x = \psi(y)‡$$

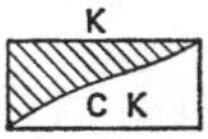

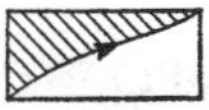

in such a manner that $R \cap K$ is defined by

$$y \geqslant \varphi(x) \quad \text{and also by} \quad x \leqslant \psi(y).$$

The preceding definition is justified by:

PROPOSITION 4.4.2. *For every* (x_0, y_0) *other than one of the angular points of the boundary* ∂K, *there exists a* K-*privileged rectangle* R *which contains* (x_0, y_0) *in its interior.*

PROOF. The result is obvious if $(x_0, y_0) \notin \partial K$; for (x_0, y_0) is the centre of a rectangle which does not meet ∂K. Suppose therefore that (x_0, y_0) is a *regular* point of the boundary ∂K; at this point the curve ∂K has an *oriented tangent* (since ∂K has been oriented). We can make a direct displacement such that, in terms of the new coordinates (x, y), this oriented tangent makes with the axis of x an angle > 0 and $< \pi/2$. In these coordinates ∂K is defined, in the neighbourhood of (x_0, y_0), by

$$y = \varphi(x), \quad \text{and also by} \quad x = \psi(y),$$

where φ and ψ are of class C^1, with $\varphi'(x_0) > 0$ and $\psi'(y_0) > 0$; if we take on ∂K, on opposite sides of (x_0, y_0), two points (a, b) and (a', b') sufficiently close, the rectangle R with vertices (a, b), (a, b'), (a', b) and (a', b') satisfies condition (b) above: it is K-privileged.

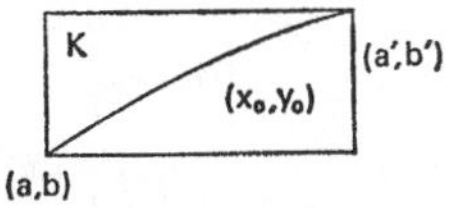

Q.E.D.

4.5 *Proof of Theorem 4.4.1 (Stokes's theorem)*

First step: the case where the support of the 1-*form* α *is contained in a* K-*privileged rectangle* R.

We can always suppose that the form α is defined and of class C^1 in the whole of the

---

† By "rectangle" we do not mean the boundary of the rectangle, but the closure of its interior; a rectangle is a particular type of a "compact set with boundary".

‡ φ and ψ being strictly increasing functions of class C^1, one being inverse of the other.

plane; if not, we can proceed as at the beginning of § 4.3. Because of property (iv) of the double integral (§ 4.3), we can suppose that the sides of R are parallel to the axes. If R does not meet the boundary ∂K (case (a)), the form α is identically zero in the neighbourhood of the boundary ∂K, therefore the integral $\int_{\partial K} \alpha$ is zero. We now want to prove that $\iint_K d\alpha = 0$, which will give us (4.4.1). To do this we shall show that

$$\iint_K \frac{\partial Q}{\partial x}\, dx \wedge dy = 0, \qquad \iint_K \frac{\partial P}{\partial y}\, dx \wedge dy = 0.$$

To prove the first of these, for example, we have by (4.3.3)

$$\iint_K \frac{\partial Q}{\partial x}\, dx \wedge dy = \iint_{K \cap R} \frac{\partial Q}{\partial x}\, dx \wedge dy,$$

since the support of Q is contained in R: *a fortiori* the support of $\partial Q/\partial x$, which is contained in supp Q, is contained in R. By (4.3.4) it follows that

$$\iint_{K \cap R} \frac{\partial Q}{\partial x}\, dx \wedge dy = \int_b^{b'} dy \left(\int_a^{a'} \frac{\partial Q}{\partial x}\, dx \right).$$

Now,

$$\int_a^{a'} \frac{\partial Q}{\partial x}\, dx = Q(a', y) - Q(a, y) = 0,$$

since the support of Q does not meet the boundary of R.

Suppose now that the K-privileged rectangle R (whose interior contains, by hypothesis, the supports of the coefficients P and Q of α), is of type (b) of § 4.4. We may suppose, as above, that the sides of R are parallel to the axes. We shall prove each of the relations

$$\iint_K \frac{\partial Q}{\partial x}\, dx \wedge dy = \int_{\partial K} Q\, dy \tag{4.5.1}$$

$$-\iint_K \frac{\partial P}{\partial y}\, dx \wedge dy = \int_{\partial K} P\, dx. \tag{4.5.2}$$

Take, for example, the first. Because of the hypothesis concerning the support of Q, it suffices to prove that

$$\iint_{K \cap R} \frac{\partial Q}{\partial x}\, dx \wedge dy = \int_{(\partial K) \cap R} Q\, dy.$$

Then, by (4.3.3), and with the notation of (b):

$$\iint_{K \cap R} \frac{\partial Q}{\partial x}\, dx \wedge dy = \int_b^{b'} dy \left(\int_a^{\psi(y)} \frac{\partial Q}{\partial x}(x, y)\, dx \right).$$

Now

$$\int_a^{\psi(y)} \frac{\partial Q}{\partial x}\, dy = Q(\psi(y), y) - Q(a, y) = Q(\psi(y), y),$$

since Q is zero on the boundary of R. Therefore,

$$\iint_{K \cap R} \frac{\partial Q}{\partial x}\, dx \wedge dy = \int_b^{b'} Q(\psi(y), y)\, dy,$$

and the right-hand side is none other than the value of the curvilinear integral

$$\int_{(\partial K)\cap R} Q\,dy$$

(we take y as the parameter, which increases from b to b'). Thus (4.5.1) is proved.

Similarly,

$$\begin{aligned} -\iint_K \frac{\partial P}{\partial y}\,dx \wedge dy &= -\iint_{K\cap R} \frac{\partial P}{\partial y}\,dx \wedge dy \\ &= -\int_a^{a'} dx\left(\int_{\varphi(x)}^{b'} \frac{\partial P}{\partial y}\,dy\right) \\ &= \int_a^{a'} P(x, \varphi(x))\,dx \end{aligned}$$

and this is equal to the curvilinear integral

$$\int_{(\partial K)\cap R} P\,dx.$$

Second step: We shall prove (4.4.1) in the case where the *support of the* 1-*form* α *contains none of the angular points* (which are of finite number) *of the boundary* ∂K.

We can always suppose that the set supp(α) is *compact* (this can be achieved by multiplying α by a numerical function of class C^∞, of compact support, and equal to 1 in a neighbourhood of K). By Prop. 4.4.2 each point of the compact set supp (α) lies in the interior of a K-privileged rectangle. Because it is compact, supp (α) can always be covered by a finite family $(U_i)_{i\in I}$, where each open U_i is the interior of a K-privileged rectangle. The U_i also cover a compact neighbourhood K′ of supp (α). By Theorem 4.1.1, there exists a differentiable partition of unity over K′, subordinate to the covering U_i. It is evident that

$$\alpha = \sum_{i\in I} (f_i\alpha).$$

Now the support of $f_i\alpha$ is contained in the interior of a privileged rectangle. We can therefore apply that which has just been proved in the "first step", i.e.,

$$\iint_K d(f_i\alpha) = \int_{\partial K} f_i\alpha.$$

Summing over $i \in I$, we obtain the relation (4.4.1).

Third and last step: the general case. In using, as above, a differentiable partition of unity, one is led to the proof of Stokes's theorem (4.4.1) in the case *where the support of* α *contains only one angular point of* ∂K. Let $z_0 = (x_0, y_0)$ be this angular point. We shall multiply α by a numerical function, zero in the neighbourhood of z_0, so that we are led to the case of the "second step". More precisely, there exists a function

$$\nu: \mathbf{R}^2 \to [0, 1],$$

of class C^∞, having the following properties:

$$\nu(z) = 0 \quad \text{for} \quad \|z\| \leqslant \tfrac{1}{2}$$
$$\nu(z) = 1 \quad \text{for} \quad \|z\| \geqslant 1$$

(where we have the used Euclidean norm). This follows immediately from Corollary 4.1.2. For each $r > 0$, put

$$\nu_r(z) = \nu\left(\frac{z - z_0}{r}\right).$$

Then the 1-form $\nu_r\alpha$ coincides with α outside the disc $\|z - z\| \leqslant r$, and is identically zero inside the disc $\|z - z_0\| \leqslant r/2$.

By what we have proved in the second step,

$$\iint_K d(\nu_r\alpha) = \int_{\partial K} \nu_r\alpha. \tag{4.5.3}$$

If we can prove that

$$\int_{\partial K} \alpha = \lim_{r\to 0} \int_{\partial K} \nu_r\alpha \tag{4.5.4}$$

and that

$$\iint_K d\alpha = \lim_{r\to 0} \iint_K d(\nu_r\alpha), \tag{4.5.5}$$

equation (4.4.1) will be obtained by passing to the limit in (4.5.3). It remains therefore, to prove (4.5.4) and (4.5.5).

We have

$$\int_{\partial K} \nu_r\alpha = \int_{\partial K} (\nu_r P)\, dx + (\nu_r Q)\, dy;$$

at each point $(x, y) \neq (x_0, y_0)$, it is evident that

$$\begin{cases} \lim\limits_{r\to 0} \nu_r(x, y)P(x, y) = P(x, y) \\ \lim\limits_{r\to 0} \nu_r(x, y)Q(x, y) = Q(x, y), \end{cases} \tag{4.5.6}$$

and further that the functions $|\nu_r P|$ and $|\nu_r Q|$ are bounded by a fixed number. Thus, Lebesgue's theorem (concerning the passage to the limit under the sign of integration) shows that (4.5.6) implies (4.5.4). For an analogous reason,

$$\iint_K d\alpha = \lim_{r\to 0} \iint_K \nu_r(d\alpha).$$

Now

$$d(\nu_r\alpha) = (d\nu_r) \wedge \alpha + \nu_r(d\alpha).$$

Therefore the equality (4.5.5) will be proved if we can show that

$$\lim_{r\to 0} \iint_K (d\nu_r) \wedge \alpha = 0. \tag{4.5.7}$$

But

$$(dv_r) \wedge \alpha = \left(Q\frac{\partial v_r}{\partial x} - P\frac{\partial v_r}{\partial y}\right) dx \wedge dy.$$

Let M be an upper bound of $\|P(x, y)\|$ and of $\|Q(x, y)\|$ on K, m_r an upper bound of $\|\partial v_r/\partial x\|$ and of $\|\partial v_r/\partial y\|$ on K. Since v_r is constant outside the disc $\|z - z_0\| \leqslant r$, we have in fact

$$\iint_K (dv_r) \wedge \alpha = \iint_{\|z-z_0\| \leqslant r} \left(Q\frac{\partial v_r}{\partial x} - P\frac{\partial v_r}{\partial y}\right) dx \wedge dy,$$

and the norm of this is less than

$$M \cdot m_r \cdot (\pi r^2),$$

since the area of a disc of radius r is πr^2. It remains to evaluate m_r: the derivatives $\partial v/\partial x$ and $\partial v/\partial y$ are continuous functions of compact support, therefore there exists $m > 0$ such that

$$\left\|\frac{\partial v}{\partial x}\right\| \leqslant m, \qquad \left\|\frac{\partial v}{\partial y}\right\| \leqslant m$$

at every point $z \in \mathbf{R}^2$. Since $v_r(z) = v((z - z_0)/r)$, we have,

$$\frac{\partial v_r}{\partial x} = \frac{1}{r}\frac{\partial v}{\partial x}, \qquad \frac{\partial v_r}{\partial y} = \frac{1}{r}\frac{\partial v}{\partial y},$$

therefore we may take

$$m_r = \frac{m}{r}.$$

Finally

$$\left\|\int_K (dv_r) \wedge \alpha\right\| \leqslant \pi M m r,$$

which proves (4.5.7). This completes the proof of Theorem 4.5.1.

4.6 *Change of variable in double integrals*

Let K be a compact set with boundary in the plane $\mathbf{R}^2$, with K piecewise of class C^1 (cf. § 4.2). If φ is a C^1-diffeomorphism of an open U contained in K onto an open U′ contained in $\mathbf{R}^2$, then $K' = \varphi(K)$ is a compact set with boundary, the boundary being piecewise of class C^1: this follows from the definitions.

Since φ is a C^1-diffeomorphism the linear mapping $\varphi'(z)$ (for $z \in U$) is an *isomorphism* $\mathbf{R}^2 \to \mathbf{R}^2$; its determinant is therefore $\neq 0$ for every $z \in U$. Recall that this determinant, called the *Jacobian* of the transformation, is

$$\frac{\partial(x', y')}{\partial(x, y)},$$

where x', y' are the coordinates of the point $\varphi(x, y)$, the image of the point z with coordinates x, y under the transformation φ. The sign of the Jacobian is locally constant:

therefore if U is *connected* (which we shall henceforth suppose to be the case, for simplicity), the Jacobian of the C^1-diffeomorphism φ has a constant sign.

DEFINITION. If the Jacobian $\varphi > 0$, we say that φ *preserves the orientation*; if the Jacobian is < 0, we say that φ *changes the orientation.*

We propose to prove the formula for the change of variable:

THEOREM 4.6.1. *With the preceding hypotheses and notations,* let ω be a differential 2-form, of class C^0, in an open U′: then

$$\iint_{\varphi(K)} \omega = \epsilon \iint_K \varphi^*(\omega) \tag{4.6.1}$$

where $\epsilon = +1$ if φ preserves the orientation, $\epsilon = -1$ if φ changes the orientation.

Let us write out the explicit form of (4.6.1): by writing in canonical form we have,

$$\omega = f(x', y')\, dx' \wedge dy',$$

x' and y' being the coordinates of a point of the open $U' = \varphi(U)$; f is continuous in U′. Then

$$\varphi^*(\omega) = f(x'(x, y), y'(x, y))\, \frac{\partial(x', y')}{\partial(x, y)}\, dx \wedge dy,$$

where $x'(x, y), y'(x, y)$ are the functions which define the C^1-diffeomorphism φ. Taking account of the value of $\epsilon = \pm 1$ in (4.6.1), the relation (4.6.1) becomes therefore

$$\iint_{\varphi(K)} f(x', y')\, dx' \wedge dy' = \iint_K f(x'(x, y), y'(x, y)) \left| \frac{\partial(x', y')}{\partial(x, y)} \right| dx \wedge dy. \tag{4.6.2}$$

This is the relation to be proved.

Now, by definition

$$\iint_{\varphi(K)} f(x', y')\, dx' \wedge dy'$$

is equal to the integral taken over the whole plane,

$$\iint \bar{f}(x'y')\, dx' \wedge dy',$$

where $\bar{f}$ denotes the function equal to f on $\varphi(K)$, and equal to 0 elsewhere. The support of $\bar{f}$ is contained in the open $\varphi(U)$, so that the function $\bar{f} \circ \varphi$ is defined (equal to $\bar{f} \circ \varphi$ in U, and equal to 0 elsewhere). Equation (4.6.2) to be established, may therefore be written

$$\iint \bar{f}\, dx' \wedge dy' = \iint \bar{f} \circ \varphi \left| \frac{\partial(x', y')}{\partial(x, y)} \right| dx \wedge dy,$$

the integral being taken over the whole plane $\mathbf{R}^2$.

Thus Theorem 4.6.1 is merely a particular case of the following (in the statement of which we write f instead of $\bar{f}$):

THEOREM 4.6.2. *Let φ denote a C^1-diffeomorphism of an open $U \subset \mathbf{R}^2$ onto an open $U' \subset \mathbf{R}^2$, and let f be a Lebesgue-integrable function, with compact support contained in U'. Then*

$$\iint_{\mathbf{R}^2} f\,dx' \wedge dy' = \iint_{\mathbf{R}^2} f \circ \varphi \left|\frac{\partial(x', y')}{\partial(x, y)}\right| dx \wedge dy. \tag{4.6.3}$$

(The fact that $\partial(x', y')/\partial(x, y)$ is only defined in U is not important, for the support of the function $f \circ \varphi$ is contained in U.)

PROOF. (a) We shall first of all prove (4.6.3) *when f is of class C^1*, and φ is a *C^2-diffeomorphism* of an open U onto an open U′ containing the compact support of f. There exists a compact set with boundary K′ containing supp f and contained in U: this can be seen by taking a sufficiently fine "net" of $\mathbf{R}^2$ (fine enough for every square of the net which meets supp f to be contained in U); it then suffices to take for K′ the union of those squares which meet supp f. Let $K = \varphi^{-1}(K')$; then

$$\iint_{\mathbf{R}^2} f\,dx' \wedge dy' = \iint_{K'} \omega, \quad \text{with} \quad \omega = f\,dx' \wedge dy',$$

$$\iint_{\mathbf{R}^2} f \circ \varphi \left|\frac{\partial(x', y')}{\partial(x, y)}\right| dx \wedge dy = \epsilon \iint_K \varphi^*(\omega),$$

so that the relation (4.6.3) to be proved is now written in the form (4.6.1). We shall therefore prove (4.6.1) in the case where ω is of class C^1, and φ a C^2-diffeomorphism.

Since ω is of class C^1, $d\omega$ exists; moreover $d\omega = 0$, since $d\omega$ is of degree 3, and $\mathbf{R}^2$ is of dimension 2. We can thus apply *Poincaré's Theorem* to ω (Theorem 2.12.1), for the plane $\mathbf{R}^2$ is starlike with respect to the origin: there exists therefore a differential 1-form α of class C^1 in $\mathbf{R}^2$, such that $d\alpha = \omega$. Therefore, by Stokes's theorem (Theorem 4.4.1)

$$\iint_{\varphi(K)} \omega = \iint_{\varphi(K)} d\alpha = \int_{\partial\varphi(K)} \alpha. \tag{4.6.4}$$

Now the formula for the change of variable for curvilinear integrals is already known (cf. § 3.3); here, the boundary $\partial\varphi(K)$ is the image, under φ, of the boundary ∂K; but we must determine whether or not the mapping φ of ∂K onto $\partial\varphi K$ preserves the orientation of the two boundaries. If $\varphi\colon U \to U'$ preserves the orientation (i.e., if the Jacobian of φ is > 0), then φ transforms the oriented boundary ∂K into the oriented boundary $\partial\varphi(K)$; on the contrary, if φ changes the orientation, φ transforms the oriented boundary ∂K into the boundary $\partial\varphi(K)$ with orientation *reversed*. This follows from the definition of the orientation of the boundary ∂K (cf. § 4.2).

From this it follows that

$$\int_{\partial\varphi(K)} \alpha = \epsilon \int_{\partial K} \varphi^*(\alpha) = \epsilon \int_{\partial K} \varphi^*(d\omega); \tag{4.6.5}$$

now (Theorem 2.9.2) $\varphi^*(d\omega) = d(\varphi^*(\omega))$, since φ being of class C^2, $\varphi^*(\omega)$ is of class C^1, therefore $d(\varphi^*(\omega))$ has a sense. Therefore (4.6.5) gives

$$\int_{\partial\varphi(K)} \alpha = \epsilon \int_{\partial K} d(\varphi^*(\omega)),$$

and by applying Stokes's theorem again we obtain

$$\int_{\partial\varphi(K)} \alpha = \epsilon \iint_K \varphi^*(\omega), \tag{4.6.6}$$

which, along with (4.6.4), gives the relation (4.6.1) to be proved.

(b) We have just proved Theorem 4.6.2 in the particular case where f is of class C^1, and φ is a C^2-diffeomorphism. The reader who is pressed for time may assume the theorem established under the general hypotheses of its statement. For the curious people we shall now give the proof in the general case. To do this we shall use the function λ, of class C^∞, defined in § 4.1; more precisely, consider the function λ in the plane $\mathbf{R}^2$:

$$\lambda(x, y) = \begin{cases} \exp\left(\dfrac{1}{x^2 + y^2 - 1}\right) & \text{for } x^2 + y^2 < 1 \\ 0 \quad \text{for } x^2 + y^2 \geqslant 1. \end{cases}$$

For each $r > 0$, let $\lambda_r(x, y) = \lambda(x/r, y/r)$, a function whose support is contained in the disc $x^2 + y^2 \leqslant r^2$; there exists a constant $c_r > 0$ such that

$$\iint c_r \lambda_r(x, y)\, dx \wedge dy = 1;$$

put $c_r\lambda_r = \mu_r$. For every function f, integrable and of compact support, we shall use the *convolution* $\mu_r * f$: this is the function defined by

$$(\mu_r * f)(x, y) = \iint f(x - u, y - v)\mu_r(u, v)\, du \wedge dv.$$

This is an integrable function of compact support, and is none other than the mean of the values of f under the translations (u, v) such that $u^2 + v^2 \leqslant r^2$. Therefore if r is small, $\mu_r * f$ is close to f, in a sense which we can make more precise: for example, if f is continuous (and therefore uniformly continuous), $\mu_r * f$ converges *uniformly* to f when r tends to 0. In the theory of Lebesgue integration, one shows that $\mu_r * f$ tends to f with respect to the norm of the space L^1: this means that

$$\iint \|f(x, y) - (\mu_r * f)(x, y)\|\, dx \wedge dy$$

tends to 0 with r; it follows that

$$\iint f(x, y)\, dx \wedge dy = \lim_{r \to 0} \iint (\mu_r * f)(x, y)\, dx \wedge dy. \tag{4.6.7}$$

On the other hand, the function $\mu_r * f$ *is of class* C^∞, as is the function μ_r: in fact, we may also write,

$$(\mu_r * f)(x, y) = \iint \mu_r(x - u, y - v) f(u, v)\, du \wedge dv,$$

and the rule for differentiating under the sign of integration (in the theory of Lebesgue integral) shows that $\mu_r * f$ has the derivative

$$(\mu_r * f) = \mu_r' * f,$$

i.e.,

$$\frac{\partial}{\partial x}(\mu_r * f) = \frac{\partial \mu_r}{\partial x} * f$$

By continuing, one can obtain successively the higher derivatives of $\mu_r * f$.

We shall also use the convolution $\mu_r * \varphi$, which is a mapping of class C^∞; it is defined, if not in the whole of the open set U, at least in the open U_r of points (x, y) such that the closed disc of centre (x, y) and radius r is contained in U. In U_r, the function $\mu_r * \varphi$: $U \to \mathbf{R}^2$ is of class C^∞, and in particular of class C^2. It is easily shown that

$$\frac{\partial}{\partial x}(\mu_r * \varphi) = \mu_r * \frac{\partial \varphi}{\partial x}$$

$$\frac{\partial}{\partial y}(\mu_r * \varphi) = \mu_r * \frac{\partial \varphi}{\partial y};$$

therefore the function $\mu_r * \varphi = \varphi_r$ and its partial derivatives converge uniformly to φ, $\partial\varphi/\partial x$ and $\partial\varphi/\partial y$ respectively. It follows that on every compact subset of U, φ_r is defined and of class C^1 for sufficiently small r, and that the Jacobian of φ_r has the sign of the Jacobian of φ at every point of the compact subset. More subtle reasoning shows that every compact set contained in U possesses an open neighbourhood $V \subset U$ such that (for r sufficiently small) φ_r is a C^2-diffeomorphism of V onto its image $\varphi_r(V)$, and that $\varphi_r(V)$ contains the support of the function $\lambda_r * f$. Therefore, for sufficiently small r, we can apply part (a), already proved:

$$\iint (\mu_r * f)\, dx' \wedge dy' = \iint (\mu_r * f) \circ \varphi_r |J(\varphi_r)|\, dx \wedge dy, \tag{4.6.8}$$

where $J(\varphi_r)$ is the Jacobian of the transformation φ_r. To finish, we apply the Lebesgue theorem on the passage to the limit under $\iint$: we shall not give the detailed justification here. Then as r tends to 0, the left-hand side of (4.6.8) tends to the left-hand side of (4.6.3), and the right-hand side of (4.6.8) tends to the right-hand side of (4.6.3). This proves (4.6.3).

4.7 *Varieties in* $\mathbf{R}^n$

PROPOSITION 4.7.1. Suppose we are given two integers $k \geqslant 1$, $p \geqslant 1$. Given a set $M \subset \mathbf{R}^n$ and a point $a \in M$ with coordinates $a_1, a_2, \ldots, a_n$, the following three properties are equivalent:

(i) There exists an open neighbourhood V of a and a C^k-diffeomorphism f of V onto an open $W \subset \mathbf{R}^n$, such that $f(a) = 0$, and $f(M \cap V)$ is the intersection of W with the plane of dimension p defined by the equations $y_{p+1} = 0, \ldots, y_n = 0$ (where the y_i are the coordinates of a point of W).

(ii) After a suitable permutation of the coordinates $x_1, \ldots, x_n$ of $\mathbf{R}^n \supset M$, there exist $n-p$ numerical functions $\varphi_{p+1}(x_1, \ldots, x_p), \ldots, \varphi_n(x_1, \ldots, x_p)$, of class C^k in a neighbourhood of the point $(a_1, \ldots, a_p)$ such that $\varphi_i(a_1, \ldots, a_p) = a_i$ (for $p + 1 \leqslant i \leqslant n$) and the points of M situated in a sufficiently small neighbourhood V of a are precisely those whose coordinates $x_1, \ldots, x_n$ satisfy the equations

$$x_i = \varphi_i(x_1, \ldots, x_p) \quad \text{for} \quad p + 1 \leqslant i \leqslant n.$$

(iii) There exists an open neighbourhood V of a in $\mathbf{R}^n$, an open neighbourhood Ω of 0 in $\mathbf{R}^p$, and a homeomorphism g of Ω onto $M \cap V$, such that g (considered as a function with values in $\mathbf{R}^n$) is of class C^k, and $g'(0) \in \mathscr{L}(\mathbf{R}^p; \mathbf{R}^n)$ is of rank p.

PROOF. Consider first our terminology: when a mapping $g: \Omega \to \mathbf{R}^n$ is of class C^1 such that $g'(t) \in \mathscr{L}(\mathbf{R}^p; \mathbf{R}^n)$ is of rank h for a point $t \in \Omega$, we say simply that *g is of rank h at the point t.* Thus (iii) says that g is of rank p at the origin 0.

In addition let us make the following note on (iii): let $t = (t_1, \ldots, t_p)$ be a point of Ω; the coordinates $x_1, \ldots, x_n$ of the point $g(t) = g(t_1, \ldots, t_p)$ are, by hypothesis, functions $g_1, \ldots, g_n$ of class C^k $(k \geqslant 1)$, and the matrix

$$\left(\frac{\partial g_i}{\partial t_j}(t_1, \ldots, t_p)\right)$$

is, by hypothesis, of rank p for $t_1 = \cdots = t_p = 0$. By continuity it is still of rank p if t belongs to an open neighbourhood Ω' of 0 $(\Omega' \subset \Omega)$. By replacing Ω by Ω', we can therefore suppose in the statement (iii) that $g'(t) \in \mathscr{L}(\mathbf{R}^p, \mathbf{R}^n)$ is of rank p for all $t \in \Omega$. In this case, we shall say that g is a *parametrization* (*of class* C^k) *of* M *in the neighbourhood of* a.

Now let us consider the proof of Prop. 4.7.1. First we shall show that (iii) implies (ii): (iii) being supposed true, we can make a permutation of the coordinates $x_1, \ldots, x_n$ such that the square matrix

$$\left(\frac{\partial g_i}{\partial t_j}(0_1, \ldots, 0)\right)_{\substack{1 \leqslant i \leqslant p \\ 1 \leqslant j \leqslant p}}$$

is of rank p. By the theorem of local inversion (*Differential Calculus*, Chap. 1, § 4.6), the transformation

$$x_i = g_i(t_1, \ldots, t_p) \qquad (1 \leqslant i \leqslant p) \tag{4.7.1}$$

is a C^k-diffeomorphism of an open neighbourhood Ω' of 0 onto an open neighbourhood of the point $(a_1, \ldots, a_p) \in \mathbf{R}^p$. Let

$$t_i = h_i(x_1, \ldots, x_p)$$

be the inverse transformation. By replacing Ω by Ω', and V by an open V′ such that $g(\Omega') = M \cap V'$, we see that the points of $M \cap V'$ are those points $(x_1, \ldots, x_n) \in V'$ such that

$$\begin{cases} x_{p+1} = g_{p+1}(h_1(x_1, \ldots, x_p), \ldots, h_p(x_1, \ldots, x_p)) \\ \cdot \quad \cdot \quad \cdot \quad \cdot \quad \cdot \quad \cdot \quad \cdot \quad \cdot \quad \cdot \quad \cdot \quad \cdot \quad \cdot \\ x_n = g_n(h_1(x_1, \ldots, x_p), \ldots, h_p(x_1, \ldots, x_p)). \end{cases}$$

The right-hand sides are functions $\varphi_{p+1}(x_1, \ldots, x_p), \ldots, \varphi_n(x_1, \ldots, x_p)$ of class C^k, which proves property (ii).

Now we show that (ii) implies (i): (ii) being supposed true, set

$$(4.7.2) \qquad \begin{cases} y_i = x_i - a_i & \text{for } 1 \leqslant i \leqslant p, \\ y_j = x_j - \varphi_j(x_1, \ldots, x_p) & \text{for } p+1 \leqslant j \leqslant n. \end{cases}$$

This transformation is a C^k-diffeomorphism of an open neighbourhood V of a in $\mathbf{R}^n$, onto a neighbourhood W of 0 in $\mathbf{R}^n$, and the inverse diffeomorphism is

$$x_i = y_i + a_i \quad \text{for} \quad 1 \leqslant i \leqslant p$$
$$x_j = y_j + \varphi_j(y_1 + a_1, \ldots, y_p + a_p) \quad \text{for} \quad p+1 \leqslant j \leqslant n.$$

The C^k-diffeomorphism (4.7.2) evidently maps $M \cap V$ onto the set of points $y \in W$ such that

$$y_{p+1} = 0, \ldots, y_n = 0,$$

which establishes (i).

Let us prove finally that (i) implies (iii): (i) being supposed true, let $k: W \to V$ be the C^k-diffeomorphism inverse to f. The intersection of W with $y_{p+1} = 0, \ldots, y_n = 0$ is an open $\Omega \subset \mathbf{R}^p$ (spanned by the coordinates $y_1, \ldots, y_p$), containing the origin $0 \in \mathbf{R}^p$; the restriction g of f to Ω is a mapping of class C^k whose image is precisely $M \cap V$. Further $g(0) = a$; as for the linear mapping $g'(0) \in \mathscr{L}(\mathbf{R}^n; \mathbf{R}^n)$, this is the restriction of the linear mapping $h'(0) \in \mathscr{L}(\mathbf{R}^n; \mathbf{R}^n)$ to the vector subspace defined by $y_{p+1} = 0, \ldots, y_n = 0$. Since $h'(0)$ is an isomorphism $\mathbf{R}^n \to \mathbf{R}^n$, the restriction to the subspace $\mathbf{R}^p$ is of rank p, and so $g'(0)$ is of rank p. Thus condition (iii) is fulfilled.

This completes the proof of Prop. 4.7.1.

DEFINITION. When a subset $M \subset \mathbf{R}^n$ and a point $a \in M$ satisfy one of the three (equivalent) conditions of Prop. 4.7.1, we say that M *is a variety of dimension p and of class C^k in the neighbourhood of a*. Note that necessarily $p \leqslant n$.

DEFINITION. We say that the subset $M \subset \mathbf{R}^n$ is *a variety of dimension p and of class* C^k if, for every point $a \in M$, M is a variety of dimension p and of class C^k in the neighbourhood of a.

For $p = 1$ we often say that M is a *curve* of class C^k. For $p = 2$, M is usually called a *surface* of class C^k. For example ($n = 3, p = 2$): in order that $M \subset \mathbf{R}^n$ be a surface of class C^k in a neighbourhood of a point $a \in M$, it is necessary and sufficient that one of the two following (equivalent) conditions be satisfied:

(ii) After a permutation of coordinates x_1, x_2, x_3, M is defined, in a neighbourhood of $a = (a_1, a_2, a_3)$, by an equation

$$x_3 = \varphi(x_1, x_2),$$

where φ is of class C^k in the neighbourhood of a_1, a_2, with $\varphi(a_1, a_2) = a_3$.

(iii) In the neighbourhood of a, M admits of a parametrization of class C^k:

$$x_i = g_i(t, t) \qquad 1 \leqslant i \leqslant 3, \qquad g_i(0, 0) = a_i,$$

with at least one of the determinants

$$\frac{\partial(g_2, g_3)}{\partial(t_1, t_2)}, \qquad \frac{\partial(g_3, g_1)}{\partial(t_1, t_2)}, \qquad \frac{\partial(g_1, g_2)}{\partial(t_1, t_2)}$$

being $\neq 0$ for (t_1, t_2) in the neighbourhood of $(0, 0)$.

Consider again the general case (p, n arbitrary). Property (iii) has led us to the notion of *parametrization* g (of class C^k) of a variety M of dimension p and of class C^k in the neighbourhood of a point $a \in M$. In property (i) we have to consider a C^k-diffeomorphism f. The composite (product) mapping $f \circ g$ is defined in the neighbourhood of 0 in $\mathbf{R}^p$; this is a bijective mapping of this neighbourhood onto a neighbourhood of 0 in the p-plane $y_{p+1} = 0, \ldots, y_n = 0$. The coordinates $y_1, \ldots, y_p$ of the point $(f \circ g)(t)$ are functions of class C^k of the parameters $t_1, \ldots, t_p$, and it is easily seen that the Jacobian of this transformation is $\neq 0$ in the neighbourhood of the origin. This transformation is a C^k-diffeomorphism.

Let us consider a second parametrization h (of class C^k): denote the parameters by $u_1, \ldots, u_p$ (in the neighbourhood of the origin); $f \circ h$ defines $y_1, \ldots, y_p$ as functions of $u_1, \ldots, u_p$ of class C^k, with Jacobian $\neq 0$. It follows that (in the neighbourhood of 0) h is the product of a C^k-diffeomorphism

$$t_i = \lambda_i(u_1, \ldots, u_p) \qquad (1 \leqslant i \leqslant p)$$

and of the parametrization g. In other words, *if there exist two parametrizations (of class* C^k*) of* M *in the neighbourhood of* a*, we can pass from one to the other by the application of a* C^k*-diffeomorphism on the parameters.*

PROPOSITION 4.7.2. Let g be a parametrization (of class C^k) of a variety M in the neighbourhood of a; suppose as usual $g(0) = a$. The image of the linear mapping

$$g'(0) : \mathbf{R}^p \to \mathbf{R}^n$$

is a vector subspace of dimension p, *which is independent of the choice of parametrization.* This is called the *tangent vector space* to the variety M (of dimension p) at the point $a \in M$, and is denoted by $T_a(M)$.

PROOF. If h is a second parametrization, then $h = g \circ \lambda$, where λ is a C^k-diffeomorphism of the parameter spaces. Thus,

$$h'(0) = g'(0) \circ \lambda'(0),$$

and since $\lambda'(0)$ is a *linear isomorphism* $\mathbf{R}^p \to \mathbf{R}^p$, the linear mappings $g'(0)$ and $h'(0)$ have the same image.

Example. Consider the surface in $\mathbf{R}^3$

$$x_3 = \varphi(x_1, x_2),$$

where φ is of class $C^k (k \geqslant 1)$ in the neighbourhood of (0, 0), with $\varphi(0, 0) = 0$. We may use x_1, x_2 as parameters of the surface: so that

$$x_1 = x_1, \qquad x_2 = x_2, \qquad x_3 = \varphi(x_1, x_2).$$

The matrix of the derived mapping is

$$\begin{pmatrix} 1 & 0 & \dfrac{\partial \varphi}{\partial x_1}(0,0) \\ 1 & 0 & \dfrac{\partial \varphi}{\partial x_2}(0,0) \end{pmatrix}$$

and the image of this mapping is the plane defined by the equation

$$x_3 = \frac{\partial\varphi}{\partial x_1}(0, 0)\cdot x_1 + \frac{\partial\varphi}{\partial x_2}(0, 0)\cdot x_2.$$

Suppose now that U is an open subset of $\mathbf{R}^n$, M a variety of dimension p (and of class C^k) in U, and ω a differential form of degree q in U. The function $\omega(x; \xi_1, \ldots, \xi_q)$ is defined for $x \in U$ and $\xi_1, \ldots, \xi_p \in \mathbf{R}^n$. We shall be particularly interested in its value for

$$(4.7.3) \qquad x \in M \quad \text{and} \quad \xi_1, \ldots, \xi_p \in T_x(M) \quad \text{(tangent space)}.$$

We shall say that ω *induces zero on the variety* M if $\omega(x; \xi_1, \ldots, \xi_p) = 0$ each time that $x, \xi_1, \ldots, \xi_p$ satisfy (4.7.3). If f is a local parametrization of M in the neighbourhood of one of its points a, the "change of variable" f defines a q-form $f^*(\omega)$ in the space of the parameters $t_1, \ldots, t_p$; *in order that ω induces zero on* M *in the neighbourhood of a, it is necessary and sufficient that the form $f^*(\omega)$ be identically zero.* (The proof of this is fairly simple, and is left as an exercise for the reader.)

4.8 *Orientation of a variety*

Suppose that $M \subset \mathbf{R}^n$, and that, in the neighbourhood of $a \in M$, M is a variety of dimension p and of class C^k ($k \geqslant 1$). By definition, two parametrizations g and g_1 of M, in the neighbourhood of a, *define the same orientation of* M (in the neighbourhood of a) if the change of parameters is defined by a C^k-diffeomorphism whose Jacobian is > 0; if on the contrary the change of parameters is defined by a C^k-diffeomorphism whose Jacobian is < 0, we say that the two parametrizations *define distinct orientations of* M. In this way we can divide all the local parametrizations of M (in the neighbourhood of a) into *two* classes; by definition, each of these defines an *orientation* of M in the neighbourhood of a.

Observe that if g is a parametrization, the image of the linear mapping $g'(0)$ is the tangent space $T_a(M)$; if, in the space $\mathbf{R}^p$ of parameters $t_1, \ldots, t_p$, we take a *direct frame* (a set of p vectors $\tau_1, \ldots, \tau_p$ whose determinant is > 0), its image under the linear mapping $g'(0)$ will be a *base* of the tangent space $T_a(M)$. It is seen that an orientation of M in the neighbourhood of a is defined by the *choice of frame* in the space $T_a(M)$; two frames define the same orientation if and only if the linear transformation of $T_a(M)$ which carries one frame into the other has determinant > 0.

Given a variety M (of dimension p, of class C^k) in $\mathbf{R}^n$, the topological space M can be covered by a set of open sets V_i (open subsets of M), each of which possesses a parametrization

$$f_i\colon \Omega_i \to V_i,$$

of the image V_i. The choice of f_i *orientates* M at every point of V_i. We say that the f_i define an *orientation* of M if, for every pair (i, j), f_i and f_j define the same orientation of M at each point of $V_i \cap V_j$.

We say that M is *orientable* if there exist local parametrizations f_i which orientate

M (as described above). A variety M, even if connected, is not always orientable: the "Möbius strip" is an example of a non-orientable surface.

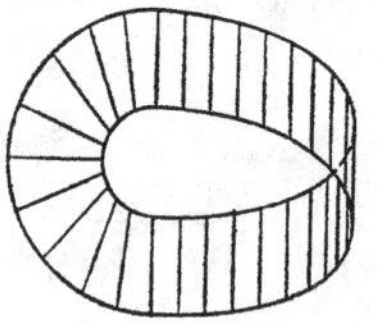

Later we shall meet examples of orientable varieties.

4.9 *Integration of a differential 2-form over a compact, oriented variety of dimension 2 and of class* C^1

Let $M \subset \mathbf{R}^n$ be a variety of dimension 2 and of class C^1. Suppose M is compact and that we are given an orientation on M. We propose to define the integral

$$\iint_M \omega \in F.$$

Let us begin by considering a *particular case*: that where $M \cap (\operatorname{supp} \omega)$ (intersection of M with the support of ω) is contained in an open connected $V \subset M$ (i.e., an open subset of a topological space M) for which there exists a parametrization of class C^1

$$\varphi: \Omega \rightarrow V,$$

Ω being an open connected subset of $\mathbf{R}^2$ (with coordinates t_1 and t_2). The parametrization φ is chosen so that it agrees with the given *orientation* on M. Observe that $\operatorname{supp} \omega \cap M$ is contained in U and is compact; therefore $\varphi(\operatorname{supp} \omega \cap M)$ is compact $\subset \Omega$. In Ω, consider the differential form $\varphi^*(\omega)$; this may be set in the form

$$f(t_1, t_2)\, dt_1 \wedge dt_2,$$

f being continuous in Ω, and of *compact support*. By definition, we set

$$\iint_M \omega = \iint_\Omega f(t_1, t_2)\, dt_1 \wedge dt_2 = \iint_\Omega \varphi^*(\omega). \tag{4.9.1}$$

To justify this definition, it must be shown that $\iint_\Omega \varphi^*(\omega)$ is *independent of the choice of the parametrization* φ.

Consider a second parametrization

$$\psi: \Omega' \rightarrow V'$$

of an open $V' \subset M$ containing the compact $M \cap (\operatorname{supp} \omega)$. Let

$$V_1 = V \cap V', \qquad \Omega_1 = \varphi^{-1}(V_1) \subset \Omega, \qquad \Omega'_1 = \psi^{-1}(V_1) \subset \Omega'.$$

Then,

$$\iint_\Omega \varphi^*(\omega) = \iint_{\Omega_1} \varphi^*(\omega) \tag{4.9.2}$$

since the support of $\varphi^*(\omega)$ is contained in Ω_1; similarly

$$\iint_{\Omega'} \psi^*(\omega) = \iint_{\Omega'_1} \psi^*(\omega). \tag{4.9.3}$$

Now, in Ω'_1, it is true that $\psi = \varphi \circ \lambda$, where λ is a C^1-diffeomorphism $\Omega'_1 \to \Omega_1$ which *preserves the orientation* (i.e., whose Jacobian is >0). Hence,

$$\psi^*(\omega) = \lambda^*(\varphi^*(\omega));$$

by the theorem concerning the change of variable in a double integral (§ 4.6), it follows that

$$\iint_{\Omega'_1} \psi^*(\omega) = \iint_{\Omega_1} \varphi^*(\omega),$$

i.e., taking account of (4.9.2) and (4.9.3):

$$\iint_{\Omega'} \psi^*(\omega) = \iint_{\Omega} \varphi^*(\omega).$$

Hence the definition (4.9.1) is justified.

If there exist two forms ω_1 and ω_2 such that $M \cap (\text{supp}\ \omega_1)$ and $M \cap (\text{supp}\ \omega_2)$ are contained in the *same* open connected $V \subset M$ for which there exists a parametrization of class C^1, then it is evident that

$$\iint_M (\omega_1 + \omega_2) = \iint_M \omega_1 + \iint_M \omega_2.$$

This remark will permit us to define $\iint_M \omega$ in the absence of the restrictive hypothesis on the support of ω.

Now consider the general case: the variety M is compact, it can be covered by a finite number of open U_i (contained in the space $\mathbf{R}^n$) such that $V_i = M \cap U_i$ admits of a parametrization. There exists a compact neighbourhood K of M which is also covered by the U_i. By Theorem 4.1.1 there exists a partition of unity (f_i) over K, subordinate to the covering (U_i); therefore

$$\begin{cases} \text{supp}\ f_i \subset U_i \\ \sum\limits_i f_i(x) = 1 \quad \text{for} \quad x \in K. \end{cases}$$

In the neighbourhood of M, the form ω is therefore the sum of the forms

$$\omega_i = f_i\omega.$$

Now $M \cap \text{supp}\ (\omega_i) \subset M \cap U_i = V_i$; therefore each ω_i satisfies the hypothesis of the "particular case" examined above. Thus the integral $\iint_M \omega_i$ is already defined. *By definition*, we set

$$\iint_M \omega = \sum_i \left(\iint_M \omega_i\right) = \sum_i \left(\iint_M f_i\omega\right). \tag{4.9.4}$$

This definition is justified if the final expression of (4.9.4) can be shown to be independent of the choice of the partition (f_i). Consider a second partition (g_j), subordinate to a covering (U'_j); let I denote the (finite) set of indices i, and J the (finite) set of indices j. If i is fixed, then $f_i(x) = \sum_{j \in J} f_i(x) g_j(x)$ when x lies in an appropriate neighbourhood of M; therefore the differential form $f_i\omega$, in the neighbourhood of M, is equal to the sum

$$\sum_{j \in J} f_i g_j \omega.$$

All the supports of these forms intersect M in a set $\subset V_i$; therefore (cf. the particular case)

$$\iint_M f_i\omega = \sum_{j \in J} \iint_M f_i g_j \omega.$$

Summing with respect to i we find that

$$\sum_{i \in I} \iint_M f_i\omega = \sum_{i \in I, j \in J} \iint_M f_i g_j \omega.$$

Similarly it is shown that

$$\sum_{j \in J} \iint_M g_j\omega = \sum_{i \in I, j \in J} \iint_M f_i g_j \omega.$$

Hence, by comparison, we obtain

$$\sum_{i \in I} \iint_M f_i\omega = \sum_{j \in J} \iint_M g_j \omega.$$

Thus the integral $\iint_M \omega$ is well defined; it is evidently a linear function of ω.

Generalization. We shall integrate on a "compact set with boundary" of a 2-dimensional oriented variety. More precisely: let $M \subset \mathbf{R}^n$ be a 2-dimensional variety of class C^1. If K is a *compact* set contained in M, we shall denote the frontier of K *in* M by ∂K (*note:* this is not the frontier of K in $\mathbf{R}^n$, which is identical to K if $n > 2$).

DEFINITION. We shall say that the compact set with boundary $K \subset M$ is of class C^1 if: (a) in $\mathbf{R}^n$, ∂K is a curve piecewise of class C^1 (of course this curve is contained in M; it has a *finite* number, possibly zero, of angular points); (b) every non-angular point a of ∂K possesses an open neighbourhood V *in* M, such that $V \cap \complement(\partial K)$ consists of two connected components: one consisting of points of $V \cap \complement K$, the other of points of V in the interior of K.

The definition resembles that already given for the particular case in which M is the plane $\mathbf{R}^2$ (cf. § 4.2). As in that case, we associate with the given orientation of M an orientation of the arcs of class C^1 of the boundary ∂K; at each non-angular point $a \in \partial K$, axes (in the tangent plane $T_a(M)$) with one axis parallel to the tangent at a (in the sense of the orientation of ∂K) and the other directed into the interior of K, form a direct frame for the given orientation of the tangent plane $T_a(M)$.

If ω is a differential 2-form in the neighbourhood of K, we define $\iint_K \omega$ by means of a partition of unity as above.

After this, Stokes's theorem (Theorem 4.4.1) may be generalized as follows:

THEOREM 4.9.1. *If* K *is a compact set with boundary of class* C^1 *in the* 2-*dimensional orientated variety* M (*of class* C^1) *and if* α *is a differential* 1-*form of class* C^1 *in the neighbourhood of* K, *then*

$$\iint_K d\alpha = \int_{\partial K} \alpha, \tag{4.9.5}$$

(with the above explained convention regarding the orientation of ∂K).

Abridged proof. By virtue of a partition of unity we are required to prove (4.9.1) in the case where the support of α intersects K in a compact set contained in an open $V \subset M$ in which there exists a parametrization $\varphi: \Omega \to V$. The left- and right-hand sides of (4.9.5) are equal respectively to

$$\iint_{\varphi^{-1}(K)} \varphi^*(d\alpha) \quad \text{and} \quad \iint_{\varphi^{-1}(\partial K)} \varphi^*(\alpha);$$

furthermore, it is true that $\varphi^*(d\alpha) = d\varphi^*(\alpha)$ (actually there is some difficulty if φ is not of class C^2, but this will not concern us here). Setting $\varphi^*(\alpha) = \beta$ (a differential form in Ω, of compact support), we are led to prove that

$$\iint_{\varphi^{-1}(K)} d\beta = \int_{\partial(\varphi^{-1}(K))} \beta,$$

which is "nearly" the Stokes formula: "nearly" because $\varphi^{-1}(K)$ is not compact; but this difficulty disappears because β has a compact support.

Example. Take $n = 3$, and denote the coordinates of $\mathbf{R}^3$ by x, y, z. Let M be a *surface* in $\mathbf{R}^3$, supposed oriented, and let $K \subset M$ be a compact set with boundary. If $\alpha = P\,dx + Q\,dy + R\,dz$ (where P, Q, R are functions of class C^1 of x, y, z in the neighbourhood of K), then equation (4.9.5) becomes

$$\iint_K \left(\frac{\partial R}{\partial y} - \frac{\partial Q}{\partial z}\right) dy \wedge dz + \left(\frac{\partial P}{\partial z} - \frac{\partial R}{\partial x}\right) dz \wedge dx + \left(\frac{\partial Q}{\partial x} - \frac{\partial P}{\partial y}\right) dx \wedge dy$$
$$= \int_{\partial K} P\,dx + Q\,dy + R\,dz$$

4.10 *Multiple integrals*

We shall limit ourselves to an outline of the theory. In the plane $\mathbf{R}^2$ we have considered compact sets K with "boundaries" piecewise of class C^1; if we suppose that ∂K has *no* angular points, then we have the notion of compact set with boundary of class C^1. We shall generalize this notion to the case of n-dimensions, and neglect the possibility of there being "angular" points, which makes the theory rather complicated.

DEFINITION. In $\mathbf{R}^n$, *a compact set with boundary of class* C^1 is a compact set K which satisfies the two following conditions:

(a) the set ∂K of frontier points is a (compact) variety of dimension $n - 1$;

(b) if $a \in \partial K$, there exists a C^1-diffeomorphism φ of an open neighbourhood V of a onto an open ball B, such that $\varphi(a) = 0$, $\varphi(K \cap V)$ is the set of points of V whose first coordinate x_1 is $\leqslant 0$, and $\varphi((\partial K) \cap V)$ is the set of points of B such that $x_1 = 0$.

For $n = 2$ this definition is indeed equivalent to that already given.

We define an *orientation* of the variety ∂K, of dimension $n - 1$, in the following manner: for each point $a \in \partial K$, choose a frame $(e_1, \ldots, e_n)$ (base of the space $\mathbf{R}^n$) such that:

(i) this reference frame is *direct* (its determinant is > 0);

(ii) e_1 is directed into the exterior of K, while $e_2, \ldots, e_n$ are *tangential* to ∂K.

Now $(e_2, \ldots, e_n)$ is a base of the tangent space $T_a(\partial K)$, which *orients* this space. The orientation of $T_a(\partial K)$ thus found is independent of the choice of the base. We see that if the diffeomorphism φ of condition (b) is chosen in such a manner that its Jacobian > 0, we obtain a C^1-diffeomorphism of the subspace $x_1 = 0$ (oriented by the coordinates $x_2, \ldots, x_n$ in this order) onto $(\partial K) \cap V$, which defines the orientation of ∂K in the neighbourhood of a.

Remark. It follows from this that the boundary of a boundary compact set is always an orientable variety.

Exercise. For $n = 2$, verify that the definition given here for the orientation of ∂K agrees with that given in § 4.2.

Notation for multiple integrals: instead of writing n integration signs $\int$, we shall write $\int^{(n)}$. Let ω be a differential n-form of class C^0 in the neighbourhood of a compact K; then

$$\omega = f(x_1, \ldots, x_n)\, dx_1 \wedge \cdots \wedge dx_n,$$

f being continuous in the neighbourhood of K. As in the case $n = 2$, we put, by definition,

$$\int_K^{(n)} \omega = \int_{\mathbf{R}^n}^{(n)} \bar{f}(x_1, \ldots, x_n)\, dx_1 \wedge \cdots \wedge dx_n$$

where $\bar{f}$ denotes that function which is equal to f on K, and to 0 elsewhere. It is a Lebesgue-integrable function, and Riemann-integrable if K is a boundary compact set. This integral possesses properties analogous to (i), (ii), (iii), (iv) stated in § 4.3. Here are two fundamental theorems:

THEOREM I_n *(change of variable). If φ is a C^1-diffeomorphism of an open connected neighbourhood* $U \supset K$ *onto an open* $U' \subset \mathbf{R}^n$, *and if ω is a differential n-form of class C^0 in U', then*

$$\int_{\varphi(K)}^{(n)} \omega = \epsilon \int_K^{(n)} \varphi^*(\omega),$$

where $\epsilon = +1$ if φ preserves the orientation (Jacobian > 0), $\epsilon = -1$ in the contrary case.

This theorem permits us to define

$$\int_M^{(n)} \omega;$$

in the case where M is an n-dimensional, oriented, compact variety, of class C^1 in a space $\mathbf{R}^p (p > n)$, and ω is an n-form in the neighbourhood of M: indeed, we proceed as in § 4.9, using a partition of unity; we see that when the support of ω intersects M in a compact set contained in a $V \subset M$ for which there exists a parametrization, then the integral

$$\int_{M}^{(n)} \omega$$

defined by this parametrization is *independent of the choice of the parametrization* (it is here that Theorem I_n is used).

THEOREM II_n (Stokes's theorem). *Let K be a compact set with boundary of class C^1, in $\mathbf{R}^n$, and let α be a differential $(n - 1)$-form of class C^1 in the neighbourhood of K. Then*

$$\boxed{\int_{K}^{(n)} d\alpha = \int_{\partial K}^{(n-1)} \alpha},$$

provided that ∂K is oriented as described above.

We shall not give the proofs of these theorems, but merely indicate the principle: suppose I_{n-1} is already proven (for example, I_1 is well known). Then we can deduce II_n, exactly as we deduced II_2 with the aid of I_1 in § 4.5. To do this we use a partition of unity, and make an explicit calculation, supposing, for example, that $\alpha = P\, dx_2 \wedge \cdots \wedge dx_n$. Next, having obtained II_n, we prove I_n, in exactly the same manner that we proved Theorem 4.6.2 using Stokes's formula in § 4.6. We shall omit the details, which are rather tedious.

Theorem II_n may be generalized further by considering the notion of compact set with boundary in a variety M (of dimension n) contained in an $\mathbf{R}^p$ $(p > n)$. We obtain a theorem which reduces to Theorem 4.9.1 in the case $n = 2$.

Example. Let us apply Theorem II_n for $n = 3$. Suppose then that K is a compact set with boundary of $\mathbf{R}^3$; ∂K is a *surface* (of 2 dimensions) of class C^1. Let α be the 2-form

$$\alpha = A\, dy \wedge dz + B\, dz \wedge dx + C\, dx \wedge dy,$$

where A, B, C are functions of x, y, z, of class C^1 in the neighbourhood of K. Then

$$\boxed{\iiint_K \left(\frac{\partial A}{\partial x} + \frac{\partial B}{\partial y} + \frac{\partial C}{\partial z}\right) dx \wedge dy \wedge dz = \iint_{\partial K} A\, dy \wedge dz + B\, dz \wedge dx + C\, dx \wedge dy}$$

(This is often referred to as *Ostrogradski's formula.*)

4.11 *Differential forms on a variety* $M \subset \mathbf{R}^n$

By definition, a differential form ω of degree q on a p-dimensional variety M of class C^k $(k \geqslant 1)$ is a function

$$\omega(x; \xi_1, \ldots, \xi_q),$$

where $x \in M$ and the vectors $\xi_1, \ldots, \xi_q$ lie in the tangent space $T_x(M)$; we suppose that, for each $x \in M$, the function

$$(\xi_1, \ldots, \xi_q) \mapsto \omega(x; \xi_1, \ldots, \xi_q)$$

is *multilinear alternating* on the vector space $T_x(M)$.

It is clear that, if α is a differential q-form in an open $U \supset M$; then a differential q-form of the above type is "induced" on M; indeed, $\alpha(x; \xi_1, \ldots, \xi_q)$ is defined for $x \in U$, $\xi_1, \ldots, \xi_q \in \mathbf{R}^n$, and is multilinear alternating in $\xi_1, \ldots, \xi_q$; ω is thus obtained by restriction of the function α.

Let V be an open set $\subset$ M, and let $\varphi: \Omega \to V$ be a parametrization of class C^k (Ω is an open subset of $\mathbf{R}^n$). If ω is a differential q-form on M, the function

$$\omega(\varphi(t); \varphi'(t)\cdot\tau_1, \ldots, \varphi'(t)\cdot\tau_q)$$

where $t \in \Omega$, and $\tau_1, \ldots, \tau_q$ are vectors of $\mathbf{R}^p$, is a differential q-form in the open $\Omega \subset \mathbf{R}^p$: this is evident. This form will be denoted by $\varphi^*(\omega)$. If $\varphi^*(\omega)$ is of class C^h (with $h < k$), then this is also the case for every other parametrization of class C^k. (*Exercise:* prove this by means of a change of parametrization.) We shall say that *the q-form ω, on* M, *is of class* C^h if every point of M possesses an open neighbourhood V (in M) having this property. Note that the *notion of a differential form of class* C^h, *on a variety* M, *has a meaning only if the variety is at least of class* C^{h+1}.

We define in an obvious manner the *sum* of two differential forms of degree q on M. An *exterior product* of two forms α and β may also be defined by means of the exterior product of two multilinear alternating forms, as in § 2.2. Similarly $d\omega$ can be defined for a form ω of class C^h ($h \geqslant 1$). We shall not develop this theory here.

Suppose now that M is a p-dimensional variety of class C^1, and let ω be a differential p-form on M, of class C^0 and of compact support. Suppose an *orientation* on M is given (which implies that M is orientable). Then, proceeding exactly as in § 4.10, we can define the integral $\int_M^{(p)} \omega$: by means of a partition of unity, we are led to consider the case where the support of ω intersects M in a compact set contained in an open $V \subset M$, and such that there exists a parametrization $\varphi: \Omega \to V$ compatible with the orientation of M. We thus put

$$\int_M^{(p)} \omega = \int_\Omega^{(p)} \varphi^*(\omega);$$

the right-hand side being independent of the choice of φ.

4.12 *The p-dimensional volume element of a variety* M *of dimension p* ($M \subset \mathbf{R}^n$)

Suppose that M is an *oriented* variety of dimension p and of class C^1. We shall define on M a differential p-form ω, of class C^0, called *the p-dimensional volume element.* In the space $\mathbf{R}^n \supset M$, we have the fundamental quadratic form (sum of the squares of the coordinates); thus the (Euclidean) *length* of a vector is defined, and the condition for two vectors to be *orthogonal* is known. Now suppose that x is a point of M; the tangent space $T_x(M)$ is a p-dimensional subspace of $\mathbf{R}^n$; choose an *orthonormal base* $(e_1, \ldots, e_p)$ of $T_x(M)$, i.e., a base consisting of mutually orthogonal unit vectors. Suppose further that the base has been chosen in such a way that the orientation of $T_x(M)$ defined by this base agrees with that already given for the variety M. (Recall that an orientation of M defines an orientation in each tangent space.) If $(e'_1, \ldots, e'_p)$ is a second orthonormal base of $T_x(M)$, compatible with the orientation, the determinant of $(e'_1, \ldots, e'_p)$ with respect to $(e_1, \ldots, e_p)$ is equal to $+1$.

Let $\xi_1, \ldots, \xi_p$ be vectors in $T_x(M)$; the determinant of these vectors with respect to the base $(e_1, \ldots, e_p)$ is independent of the choice of the base. It will be denoted by $\det(\xi_1, \ldots, \xi_p)$. Note that if the orientation of M is changed, then this determinant is multiplied by -1. We can thus define ω, the required differential form of degree p: put

$$\omega(x; \xi_1, \ldots, \xi_p) = \det(\xi_1, \ldots, \xi_p) \tag{4.12.1}$$

for $\xi_1, \ldots, \xi_p \in T_x(M)$. It can be verified that it is of class C^0, i.e., for a parametrization φ of class C^1, the coefficients of $\varphi^*(\omega)$ are *continuous*. To calculate $\det(\xi_1, \ldots, \xi_p)$ we may proceed as follows: choose an orthonormal base of the subspace of $\mathbf{R}^n$ orthogonal to $T_x(M)$, $(e_{p+1}, \ldots, e_n)$, in such a way that the determinant of

$$(e_1, \ldots, e_p, e_{p+1}, \ldots, e_n)$$

is equal to $+1$ with respect to the canonical base of $\mathbf{R}^n$. Then

$$\omega(x; \xi_1, \ldots, \xi_p) = \det(\xi_1, \ldots, \xi_p, e_{p+1}, \ldots, e_n), \tag{4.12.2}$$

the determinant being taken with respect to the canonical base of $\mathbf{R}^n$.

Volume of an open V *relatively compact in* M: this, by definition, is the integral

$$\int_V^{(p)} \omega$$

where ω denotes the p-dimensional volume element. If V admits of a parametrization φ, the volume is, by definition,

$$\int_\Omega^{(p)} \varphi^*(\omega)$$

where $\varphi^*(\omega)$ is written in the form $f(t_1, \ldots, t_p)\, dt_1 \wedge \cdots \wedge dt_p$; it is easily verified that $f(t_1, \ldots, t_p) > 0$, in view of the definitions. (This will also be clear from the following examples.) Thus, the p-dimensional volume of an open, non-empty V is > 0.

Example. $n = 3$, $p = 2$. In this case we are dealing with an oriented *surface* M in the space $\mathbf{R}^3$; let x, y, z be the coordinates of $\mathbf{R}^3$. In the neighbourhood of each point of M, there exists a parametrization φ defined by three functions

$$x = x(u, v), \qquad y = y(u, v), \qquad z = z(u, v)$$

of the real variables u, v of class C^1. The fact that φ is of rank 2 is expressed by the condition that the matrix

$$\begin{pmatrix} \dfrac{\partial x}{\partial u} & \dfrac{\partial y}{\partial u} & \dfrac{\partial z}{\partial u} \\[2ex] \dfrac{\partial x}{\partial v} & \dfrac{\partial y}{\partial v} & \dfrac{\partial z}{\partial v} \end{pmatrix} \tag{4.12.3}$$

is of rank 2. Thus at least one of the three quantities

$$p = \frac{\partial(y, z)}{\partial(u, v)}, \qquad q = \frac{\partial(z, x)}{\partial(u, v)}, \qquad r = \frac{\partial(x, y)}{\partial(u, v)}$$

is $\neq 0$ at each point of the open Ω of the (u, v)-plane.

At the point $(x, y, z) \in \mathrm{M}$, corresponding to (u, v), the *unit vector normal to* M is the vector e with components

$$\frac{p}{\sqrt{p^2 + q^2 + r^2}}, \qquad \frac{q}{\sqrt{p^2 + q^2 + r^2}}, \qquad \frac{r}{\sqrt{p^2 + q^2 + r^2}},$$

Indeed, *a priori*, it is $\neq e$; the sign ϵ must be chosen in such a manner that the reference frame formed by the two vectors (4.12.3) and e_3, is direct, i.e., its determinant

$$\begin{vmatrix} \dfrac{\partial x}{\partial u} & \dfrac{\partial y}{\partial u} & \dfrac{\partial z}{\partial u} \\[2ex] \dfrac{\partial x}{\partial v} & \dfrac{\partial y}{\partial v} & \dfrac{\partial z}{\partial v} \\[2ex] \dfrac{\epsilon p}{\sqrt{p^2 + q^2 + r^2}} & \dfrac{\epsilon q}{\sqrt{p^2 + q^2 + r^2}} & \dfrac{\epsilon r}{\sqrt{p^2 + q^2 + r^2}} \end{vmatrix}$$

is > 0. The determinant is

$$\epsilon \frac{p^2 + q^2 + r^2}{\sqrt{p^2 + q^2 + r^2}} = \epsilon\sqrt{p^2 + q^2 + r^2}.$$

We must therefore take $\epsilon = +1$, and the 2-*dimensional volume* (*or element of area*) satisfies

(4.12.4) $$\varphi^*(\omega) = \sqrt{p^2 + q^2 + r^2}\, du \wedge dv$$

for the parametrization φ.

PROPOSITION 4.12.1. *If, at each point of a surface* M, *the components of the unit normal vector to* M (*oriented*) *are denoted by* $\cos \alpha$, $\cos \beta$, $\cos \gamma$, *then the form* ω *which gives the element of area is equal to*

(4.12.5) $$\cos \alpha\, dy \wedge dz + \cos \beta\, dz \wedge dx + \cos \gamma\, dx \wedge dy.$$

PROOF. This follows from the fact that

$$dy \wedge dz = p\, du \wedge dv, \qquad dz \wedge dx = q\, du \wedge dv, \qquad dx \wedge dy = r\, du \wedge dv.$$

Second example. $n = 3$, $p = 1$. Here we consider an *oriented curve* C *in* $\mathbf{R}^3$; the 1-dimensional element of volume is called the *length element* of the curve: it is a differential 1-form on the curve. We shall show that if the coordinates x, y, z of a point of the curve are functions of class C^1: $x(t)$, $y(t)$, $z(t)$, whose derivatives $x'(t)$, $y'(t)$, $z'(t)$ are not simultaneously zero, the length element is

$$\sqrt{x'^2 + y'^2 + z'^2}\, dt,$$

provided that the orientation of the curve corresponds to increasing t.

If now the curve C is contained in a surface M (with parameters u and v), C is defined by taking u and v to be functions of t, and the length element is

$$\sqrt{\mathrm{E}u'^2 + 2\mathrm{F}u'v' + \mathrm{G}v'^2}\, dt,$$

with

$$\begin{cases} E = \left(\frac{\partial x}{\partial u}\right)^2 + \left(\frac{\partial y}{\partial u}\right)^2 + \left(\frac{\partial z}{\partial u}\right)^2 \\ F = \frac{\partial x}{\partial u}\cdot\frac{\partial x}{\partial v} + \frac{\partial y}{\partial u}\cdot\frac{\partial y}{\partial v} + \frac{\partial z}{\partial u}\cdot\frac{\partial z}{\partial v} \\ G = \left(\frac{\partial x}{\partial v}\right)^2 + \left(\frac{\partial y}{\partial v}\right)^2 + \left(\frac{\partial z}{\partial v}\right)^2. \end{cases}$$

Note that knowledge of the functions E, F, G enables us to calculate the area element $\sqrt{p^2 + q^2 + r^2}\, du \wedge dv$, since

$$\boxed{p^2 + q^2 + r^2 = EG - F^2}; \tag{4.12.6}$$

which follows from the Lagrange identity

$$(bc' - cb')^2 + (ca' - ac')^2 + (ab' - ba')^2 \\ = (a^2 + b^2 + c^2)(a'^2 + b'^2 + c'^2) - (aa' + bb' + cc')^2.$$

Thus knowledge of the length element for curves lying in the surface determines also the area element of the surface (except of course, for the sign).

5. Maxima and minima of a function on a variety

Here we return to the problem of Chap. 1, § 8 of *Differential Calculus*, which concerns the relative minimum of a real function f, defined in an open subset U of a Banach space E. We shall suppose that $E = \mathbf{R}^n$.

5.1 *First order conditions*

Let M be a variety of class C^1 in $\mathbf{R}^n$, and let $f: U \to \mathbf{R}$ be a function of class C^1 defined in an open U containing M. Consider the function $g: M \to \mathbf{R}$ obtained by *restriction* of f, and let us determine whether, at a point $a \in M$, the function g has a *relative minimum*, i.e., if

$$f(x) \geqslant f(a)$$

for all $x \in M$ sufficiently close to a. The function g possesses a *strictly relative minimum* if there exists a neighbourhood V of a such that

$$f(x) > f(a)$$

for every $x \in M \cap V$ such that $x \neq a$.

We shall give three propositions, analogous respectively to Prop. 8.1.1, Theorem 8.2.1, and Theorem 8.3.3, of Chap. 1 of *Differential Calculus*.

PROPOSITION 5.1.1. *If the restriction g of f to M possesses a relative minimum at the point $a \in M$, then the derivative*

$$f'(a) \in \mathscr{L}(\mathbf{R}^n; \mathbf{R})$$

vanishes on the tangent space $T_a(M) \subset \mathbf{R}^n$ (the *necessary* condition for a relative minimum).

PROOF. Let $\varphi: \Omega \to M$ be a parametrization of class C^1 of a neighbourhood of a in M, such that $\varphi(0) = a$. We require $f \circ \varphi$ to have a relative minimum at the point 0. If this is the case, it follows that (*Differential Calculus*, Chap. 1, Prop. 8.1.1)

$$(f \circ \varphi)'(0) = 0,$$

i.e.,

$$f'(a) \circ \varphi'(0) = 0.$$

This shows that $f'(a)$ vanishes on the image of $\varphi'(0)$, which is precisely the tangent space $T_a(M)$.

Q.E.D.

5.2 *Second order conditions*

In the following we shall suppose that M is a variety of class C^2, and that f is a function of class C^2. The bilinear form $f''(a)$ induces a symmetric bilinear mapping

$$T_a(M) \times T_a(M) \to \mathbf{R}.$$

But we shall consider a second bilinear symmetric form

$$\Phi: T_a(M) \times T_a(M) \to \mathbf{R}.$$

To define this, recall that if φ is a parametrization $\Omega \to M$ of class C^1, such that $\varphi(0) = a$ (where Ω is an open subset of $\mathbf{R}^p$ containing the origin), then $\varphi'(0)$ is a linear bijection of $\mathbf{R}^p$ onto the tangent space $T_a(M)$. Suppose that φ is of class C^2, and define Φ by the condition

(5.2.1) $\quad \Phi(\varphi'(0)\cdot\tau_1, \varphi'(0)\cdot\tau_2) = (f \circ \varphi)''(0)\cdot(\tau_1, \tau_2) \quad \text{for} \quad \tau_1 \text{ and } \tau_2 \in \mathbf{R}^p.$

We shall show that the Φ thus defined is *independent of the choice of parametrization* φ. Indeed, suppose that φ is replaced by $\psi = \varphi \circ \lambda$, where λ is a C^2-diffeomorphism of a neighbourhood of 0 (in $\mathbf{R}^p$) onto a neighbourhood of 0 (in $\mathbf{R}^p$). Then $\psi'(0) = \varphi'(0) \circ \lambda'(0)$; we must therefore show that

(5.2.2) $\quad \Phi(\varphi'(0) \circ \lambda'(0)\cdot\tau_1, \varphi'(0) \circ \lambda'(0)\cdot\tau_2) = (f \circ \varphi\cdot\lambda)''(0)\cdot(\tau_1, \tau_2).$

To simplify matters put $f \circ \varphi = h$; then, by formula (7.5.1) of Chap. 1 of *Differential Calculus*, which gives the second derivative of a composite function:

$$(h \circ \lambda)''(0)\cdot(\tau_1, \tau_2) = h''(0)\cdot(\lambda'(0)\cdot\tau_1, \lambda'(0)\cdot\tau_2) + h'(0)\cdot(\lambda''(0)\cdot(\tau_1, \tau_2)).$$

But $h'(0) = 0$, since $f \circ \varphi = h$ has a relative minimum at the origin; and so the right-hand side of (5.2.2) is equal to

$$h''(0)\cdot(\lambda'(0)\cdot\tau_1, \lambda'(0)\cdot\tau_2);$$

now, by equation (5.2.1), with τ_1 replaced by $\lambda'(0)\cdot\tau_1$ and τ_2 by $\lambda'(0)\cdot\tau_2$ this is finally equal to the right-hand side of (5.2.2). Therefore (5.2.2) is proved, and the bilinear form Φ is independent of the choice of the parametrization φ. Observe that, if formula (7.5.1) of Chap. 1 of *Differential Calculus* is applied to the right-hand side of (5.2.1) we find that, for $\xi_1 \in T_a(M)$, $\xi_2 \in T_a(M)$,

$$\Phi(\xi_1, \xi_2) = f''(a)\cdot(\xi_1, \xi_2) + f'(a)\cdot(\varphi''(0)\cdot(\varphi'(0)^{-1}\xi_1, \varphi'(0)^{-1}\xi_2)). \tag{5.2.3}$$

Thus we see that it is *important not to confuse* Φ *with the bilinear form induced on* $T_a(M)$ *by* $f''(a)$.

PROPOSITION 5.2.1. *In order that the restriction of f to the variety M possess a relative minimum at the point $a \in M$, it is necessary that*

$$\Phi(\xi, \xi) \geqslant 0 \quad \textit{for every} \quad \xi \in T_a(M).$$

PROOF. This condition is equivalent to saying that

$$(f \circ \varphi)''(0)\cdot(\tau, \tau) \geqslant 0 \quad \text{for every} \quad \tau \in \mathbf{R}^p,$$

and we know (*Differential Calculus*, Chap. 1, Theorem 8.2.1) that this condition is *necessary* if $f \circ \varphi$ is to possess a relative minimum at the origin.

Q.E.D.

PROPOSITION 5.2.2. *In order that the restriction of f to the variety M possess a strict relative minimum at the point $a \subset M$, it is sufficient that*

$$\Phi(\xi, \xi) > 0 \quad \text{for every non-null } \xi \text{ of } T_a(M).$$

PROOF. This condition says that the bilinear form $(f \circ \varphi)''(0)$ is *positive and non-degenerate;* now we know (*Differential Calculus*, Chap. 1, Theorem 8.3.3) that this is a sufficient condition for $f \circ \varphi$ to possess a *strict* relative minimum at the origin.

Q.E.D.

6. Theorem of Frobenius

6.1 *Introduction to the problem*

In Chap. 2 of *Differential Calculus* a study was made of a differential equation of the form

$$\frac{dx}{dt} = f(t, x),$$

where t is a real variable, and x is an unknown function of t taking values in a Banach space E.

We shall now replace the variable $t \in \mathbf{R}$ by a variable $x \in E$ (Banach); y will be an unknown function of x with values in a Banach space F. Thus we shall consider the "differential equation"

$$\frac{dy}{dx} = f(x, y). \tag{6.1.1}$$

Since the derivative dy/dx of a function $y(x)$ takes values in $\mathscr{L}(E; F)$, the given function $f(x, y)$ will take values in $\mathscr{L}(E; F)$. More precisely: let U be an open set $\subset E \times F$, and let

$$f\colon U \to \mathscr{L}(E; F)$$

be a given function of class C^1. A function $y = \varphi(x)$ of class C^1 in an open $V \subset E$, with values in F, and such that

(i) $(x, \varphi(x)) \in U$ for every $x \in V$,
(ii) $\varphi'(x) = f(x, \varphi(x))$ for every $x \in V$,

is called a *solution* of equation (6.1.1).

Example. Take $F = \mathbf{R}$, $E = \mathbf{R}^n$: equation (6.1.1) takes the form

$$\frac{\partial y}{\partial x_i} = f_i(x_1, \ldots, x_n, y), \qquad 1 \leqslant i \leqslant n \tag{6.1.2}$$

where the n functions f_i, are given scalar functions, of class C^1 in an open $U \subset \mathbf{R}^{n+1}$. A solution is a scalar function $\varphi(x_1, \ldots, x_n)$ such that

$$d\varphi = \sum_{i=1}^{n} f_i(x_1, \ldots, x_n, \varphi)\, dx_i.$$

Equation (6.1.2) is often written in the form

$$dy = \sum_{i=1}^{n} f_i(x_1, \ldots, x_n, y)\, dx_i. \tag{6.1.3}$$

The solutions are those functions $y(x_1, \ldots, x_n)$ which annihilate the differential form $dy - \sum_{i=1}^{n} f_i\, dx_i$. An equation such as (6.1.3) is often called a "total differential equation".

We return to the general case of equation (6.1.1). The unknown function $y = \varphi(x)$ must annihilate the differential form

$$dy - f(x, y) \cdot dx. \tag{6.1.4}$$

More precisely, this is a differential form of degree 1 in the open $U \subset E \times F$, with values in F: at the point $(x, y) \in U$ and at the vector $(\xi, \eta) \subset E \times F$, it generates the vector

$$\eta - f(x, y) \cdot \xi \in F.$$

By this "change of variable"

$$x = x, \qquad y = \varphi(x), \tag{6.1.5}$$

it defines a differential form

$$\varphi'(x) \cdot dx - f(x, \varphi(x)) \cdot dx$$

in an open $V \subset E$, with values in F. To say that $y = \varphi(x)$ is a solution of (6.1.1), is to say that the differential form obtained from (6.1.4) by the change of variable (6.1.5) is *zero*.

This suggests a generalization of the notion of "solution" of the equation (6.1.1): this will be a system of two functions $x(t)$, $y(t)$ of a variable t (which varies in an open W of Banach space G) such that the change of variable

$$x = x(t), \qquad y = y(t)$$

gives zero for the differential form (6.1.4); in other words:

$$\frac{dy}{dt} = f(x(t), y(t)) \circ \frac{dx}{dt} \tag{6.1.6}$$

[equality of two functions with values in $\mathscr{L}(G; F)$]. In particular, if t is a real variable we shall say that

$$x = x(t), \qquad y = y(t)$$

is an *integral curve* of equation (6.1.1) if these functions satisfy (6.1.6) (equality of two functions with values in F).

6.2 *The first existence theorem*

With the preceding notation, suppose we are given a point $(x_0, y_0) \in U$. We shall enquire into the existence, in an open neighbourhood V of $x_0 \in E$, of a solution $y = \varphi(x)$ such that $y_0 = \varphi(x_0)$ (solution with initial value y given at $x = x_0$). *We shall see that in general, such a solution does not exist.*

THEOREM 6.2.1. *For a given point $(x_0, y_0) \in U$, there exists a sufficiently small $r > 0$ such that, in the ball $\|x - x_0\| < r$, there exists a unique function $\varphi(x)$ of class C^1, with values in F, with the property that*

$$\varphi(x_0) = y_0 \tag{6.2.1}$$

$$\boxed{\varphi'(x)\cdot(x - x_0) = f(x, \varphi(x))\cdot(x - x_0)} \quad \text{for all } x. \tag{6.2.2}$$

Note. The condition (6.2.2) is much weaker than the condition $\varphi'(x) = f(x, \varphi(x))$ which expresses the fact that φ is a solution of (6.1.1). The condition (6.2.2) expresses only that, at the point x, $\varphi'(x)$ and $f(x, \varphi(x))$ are two elements of $\mathscr{L}(E; F)$ which take the same value on vectors of E *proportional to* $x - x_0$; it does not follow that they take the same value on *all* vectors of E, which would mean that φ is a solution of (6.1.1).

We shall call a function φ satisfying (6.2.2) a *pseudo-solution* of equation (6.1.1). Theorem (6.2.1) shows that, for a given initial value (x_0, y_0), there exists one, and only one, pseudo-solution in some neighbourhood of x_0.

PROOF OF THEOREM 6.2.1. Choose a non-zero vector $\xi \in E$, and consider, in E, the straight line

$$x = x_0 + t\xi, \qquad t \text{ varying in } \mathbf{R}.$$

Let us search for an "integral curve" (cf. end of 6.1) of the form

$$x = x_0 + t\xi, \qquad y = \psi(t), \tag{6.2.3}$$

with $\psi(0) = y$. Then $dx/dt = \xi$, and so the condition which expresses that (6.2.3) is an integral curve is

$$\psi'(t) = f(x_0 + t\xi, \psi(t))\cdot\xi. \tag{6.2.4}$$

In other words, $y = \psi(t)$ (for t in the neighbourhood of 0) is *the* solution of the (ordinary) differential equation

(6.2.5) $$\frac{dy}{dt} = f(x_0 + t\xi, y)\cdot\xi$$

which is equal to y_0 for $t = 0$.

As soon as $\|\xi\| < a$ (for sufficiently small $a > 0$) it is certain that the solution $\psi(t)$ exists in all of the interval $-1 \leqslant t \leqslant +1$. Indeed, since f is of class C^1, there exist numbers $\rho > 0$, $r > 0$, k and $M > 0$ such that

$$\|f(x,y)\| \leqslant M \quad\text{and}\quad \|f'_y(x,y)\| \leqslant k \quad\text{for}\quad \|x - x_0\| \leqslant r, \qquad \|y - y_0\| \leqslant \rho,$$

and we may suppose that $\rho \leqslant Mr$ (we have only to choose a smaller ρ). Take $a = \rho/M$; for $\|\xi\| \leqslant a$ we have

$$\|f(x_0 + t\xi, y)\cdot\xi\| \leqslant Ma \quad\text{for}\quad |t| \leqslant 1, \qquad \|y - y_0\| \leqslant \rho;$$

we are therefore assured of the existence of the solution $y = \psi(t)$ (such that $\psi(0) = y_0$) in the interval $|t| \leqslant \rho/(Ma) = 1$.

In the following let $\psi(t, \xi)$ denote the solution of (6.2.5) which is equal to y_0 for $t = 0$; it depends on $\xi \in E$. Since $f(x_0 + t\xi, y)\cdot\xi$ is of class C^1 at (t, ξ, y), it follows that $\psi(t, \xi)$ is of class C^1 (*Differential Calculus*, Chap. 2, Theorem 3.6.1). Further, if t is replaced by λt and ξ by $(1/\lambda)\xi$ (λ real, $\neq 0$), then equation (6.2.5) is unchanged; therefore $\psi(t, \xi)$ *depends only on the product* $t\xi$.

$$\psi(t, \xi) = \psi(1, t\xi).$$

For $x = x_0 + t\xi$ (ξ fixed, t variable), equation (6.2.2) is equivalent to (for $t \neq 0$)

$$\varphi'(x_0 + t\xi)\cdot\xi = f(x_0 + t\xi, \varphi(x_0 + t\xi))\cdot\xi,$$

and comparing with (6.2.4) we see that

$$\varphi(x_0 + t\xi) = \psi(1, t\xi).$$

In other words, we must have

(6.2.6) $$\boxed{\varphi(x) = \psi(1, x - x_0)},$$

and this is sufficient. This determines φ, and establishes Theorem 6.2.1.

6.3 *The second existence theorem*

It remains to resolve the following problem: under what conditions does *the* pseudo-solution of the differential equation (6.1.1) defined in Theorem 6.2.1 become a true solution? In the latter case we should have

$$\varphi'(x)\cdot\eta = f(x, \varphi(x))\cdot\eta$$

for all $\eta \in E$, and all x sufficiently close to x_0.

THEOREM 6.3.1. *In order that the pseudo-solution φ be a true solution, it is necessary and sufficient* *that the function f satisfy the following condition: for every x (in the neighbourhood of x_0) the element*

$$f'_x(x, \varphi(x)) + f'_y(x, \varphi(x)) \circ f(x, \varphi(x)) \in \mathscr{L}(E; \mathscr{L}(E; F))$$

defines a symmetric element of $\mathscr{L}_2(E; F)$ (in the bijective canonical correspondence between $\mathscr{L}(E; \mathscr{L}(E; F))$ and $\mathscr{L}_2(E; F)$).

In other words, for arbitrary ξ and $\eta \in E$, we must have

$$((f'_x + f'_y \circ f) \cdot \xi) \cdot \eta = ((f'_x + f'_y \circ f) \cdot \eta) \cdot \xi \tag{6.3.1}$$

(with the understanding that f, f'_x, f'_y are to be evaluated at the point $(x, \varphi(x))$).

Note. At each point $(x, \varphi(x))$, the value of f'_x is in $\mathscr{L}(E; \mathscr{L}(E; F))$ that of f is in $\mathscr{L}(E; F)$, that of f'_y in $\mathscr{L}(E; \mathscr{L}(E; F))$, and that of $f'_y \circ f$ is in $\mathscr{L}(E; (E; F))$ by composition:

$$E \xrightarrow{f(x, \varphi(x))} F \xrightarrow{f'_y(x, \varphi(x))} \mathscr{L}(E; F).$$

Since the value of $f'_x + f'_y \circ f$, at the point $(x, \varphi(x))$, is an element of $\mathscr{L}(E; \mathscr{L}(E; F))$, the value of

$$(f'_x + f'_y \circ f) \cdot \xi$$

is in $\mathscr{L}(E; F)$; we may therefore evaluate this for a vector $\eta \in E$, to obtain an element of F: the left-hand side of (6.3.1). The right-hand side may be interpreted in the same way.

PROOF OF THEOREM 6.3.1. (1) The condition (6.3.1) is *necessary in order that φ be a solution.* Indeed, suppose that

$$\varphi'(x) = f(x, \varphi(x)).$$

since $f(x, y)$ and $\varphi(x)$ are both of class C^1, this relation shows that φ is of class C^2, and that

$$\varphi''(x) \cdot \xi = (f'_x + f'_y \circ \varphi) \cdot \xi,$$

hence, replacing $\varphi'(x)$ by $f(x, \varphi(x))$:

$$\varphi''(x) \cdot \xi = (f'_x + f'_y \circ f) \cdot \xi,$$

and so

$$(\varphi''(x) \cdot \xi) \cdot \eta = ((f'_x + f'_y \circ f) \cdot \xi) \cdot \eta.$$

Now the left-hand side of this equality is *symmetric* in ξ and η (*Differential Calculus*, Chap. 1, Theorem 5.1.1). Therefore the right-hand side is symmetric; and we obtain (6.3.1).

Q.E.D.

The following section deals with the proof of *sufficiency* of equation (6.3.1).

6.4 *Completion of the proof of the second existence theorem* (Theorem 6.3.1)

It remains to show: (2) *the condition* (6.3.1) *is sufficient for the pseudo-solution $\varphi(x)$ to be a true solution.*

Recall that, by (6.2.2), we have

$$\varphi'(x_0 + t\xi) \cdot \xi = f(x_0 + t\xi, \varphi(x_0 + t\xi)) \cdot \xi \tag{6.4.1}$$

for every $\xi \in E$ such that $\|\xi\| \leqslant a$, and every $t \in \mathbf{R}$ such that $|t| \leqslant 1$. It is required to show that, under the same conditions on ξ and t,

$$\varphi'(x_0 + t\xi)\cdot\eta = f(x_0 + t\xi, \varphi(x_0 + t\xi))\cdot\eta \tag{6.4.2}$$

for an arbitrary vector $\eta \in E$. Fix ξ and η, and consider the function

$$H(t) = t\varphi'(x_0 + t\xi)\cdot\eta - tf(x_0 + t\xi, \varphi(x_0 + t\xi))\cdot\eta, \quad \text{for} \quad |t| \leqslant 1.$$

This is a function taking values in the Banach space F. It suffices to show that $H(t) = 0$ for every t: indeed, for $t \neq 0$ the equation $H(t) = 0$ implies (6.4.2), which is therefore true for $t \neq 0$, and so also for $t = 0$ by continuity.

To prove that $H(t) = 0$, we shall show that the function $H(t)$ satisfies a linear homogeneous differential equation; since $H(0) = 0$, it will follow that $H(t)$ is identically zero.

First let us show that $H(t)$ possesses a derivative $H'(t)$. To do this recall that (cf. § 6.2)

$$\varphi(x_0 + t\xi) = \psi(t, \xi),$$

where ψ (considered as a function of t) is a solution of the differential equation (6.2.4):

$$\frac{dy}{dt} = f(x_0 + t\xi, y)\cdot\xi,$$

which is of class C^1 in ξ. The solution $\psi(t, \xi)$ is therefore a function of class C^1 of the parameter ξ, and $\partial\psi/\partial\xi$ has a derivative with respect to t, such that

$$\frac{\partial}{\partial t}\frac{\partial\psi}{\partial\xi} = \frac{\partial}{\partial\xi}\frac{\partial\psi}{\partial t} = \frac{\partial}{\partial\xi}[f(x_0 + t\xi, \varphi(x_0 + t\xi))\cdot\xi]$$

(cf. *Differential Calculus*, Chap. 2, Theorem 3.6.1). Now it is true that

$$H(t) = \frac{\partial\psi}{\partial\xi}\cdot\eta - tf(x_0 + t\xi, \varphi(x_0 + t\xi))\cdot\eta; \tag{6.4.3}$$

therefore dH/dt exists and is equal to:

$$\begin{aligned} \frac{dH}{dt} = {} & \frac{\partial}{\partial\xi}[f(x_0 + t\xi, \varphi(x_0 + t\xi))\cdot\xi]\cdot\eta \\ & - f(x_0 + t\xi, \varphi(x_0 + t\xi))\cdot\eta \\ & - t((f'_x + f'_y \circ \varphi')\cdot\xi)\cdot\eta, \end{aligned} \tag{6.4.4}$$

with the understanding that f'_x denotes $f'_x(x_0 + t\xi, \varphi(x_0 + t\xi))$, and that f'_y denotes $f'_y(x_0 + t\xi, \varphi(x_0 + t\xi))$, and that φ' denotes $\varphi'(x_0 + t\xi)$. On the right-hand side of (6.4.4), $(f'_y \circ \varphi')\cdot\xi$ can be replaced by $(f'_y \circ f)\cdot\xi$, in virtue of (6.4.1). On the other hand, by the formula which gives the derivative of a bilinear function, we have

$$\begin{aligned} & \frac{\partial}{\partial\xi}[f(x_0 + t\xi, \varphi(x_0 + t\xi))\cdot\xi]\cdot\eta \\ & \qquad = \frac{\partial}{\partial\xi}[f(x_0 + t\xi, \varphi(x_0 + t\xi))\cdot\eta]\cdot\xi + f(x_0 + t\xi, \varphi(x_0 + t\xi))\cdot\eta. \end{aligned}$$

Substituting in (6.4.4), we find, after some simplification:

$$\frac{dH}{dt} = \frac{\partial}{\partial \xi}[f(\ldots)\cdot\eta]\cdot\xi - t((f'_x + f'_y \circ f)\cdot\xi)\cdot\eta. \tag{6.4.5}$$

Explicitly:

$$\frac{\partial}{\partial \xi}[f(x_0 + t\xi, \varphi(x_0 + t\xi))\cdot\eta] = t(f'_x + f'_y \circ \varphi')\cdot\eta;$$

on the other hand, *by hypothesis* (see the statement of Theorem 6.3.1):

$$((f'_x + f'_y \circ f)\cdot\xi)\cdot\eta = ((f'_x + f'_y \circ f)\cdot\eta)\cdot\xi.$$

Finally:

$$\frac{dH}{dt} = t[(f'_y \circ (\varphi' - f))\cdot\eta]\cdot\xi = [(f'_y \circ (t\varphi' - tf))\cdot\eta]\cdot\xi, \tag{6.4.6}$$

and since $(t\varphi' - tf)\cdot\eta = H(t)$ by definition, we obtain

$$\boxed{\frac{dH}{dt} = (f'_y \circ H)\cdot\xi}. \tag{6.4.7}$$

The right-hand side of (6.4.7) is evidently a linear continuous function of H (recall that the values of H lie in a Banach space F; those of $f'_y \circ H$ lie in $\mathscr{L}(E; F)$, as do those of f; the right-hand side of (6.4.7) is that element of F with the value $f'_y \circ H$ on the vector $\xi \in E$).

Thus $H(t)$ satisfies a linear homogeneous differential equation (6.4.7), as stated, with $H(0) = 0$. We conclude that $H(t)$ vanishes identically.

This completes the proof of Theorem 6.3.1.

6.5 *The fundamental theorem*

This is an immediate consequence of Theorem 6.3.1. With the hypotheses of § 6.1, consider the differential equation

$$\frac{dy}{dx} = f(x, y), \tag{6.5.1}$$

where $f: U \to \mathscr{L}(E; F)$ is a given function of class C^1. We now require that, *for arbitrary* $(x_0, y_0) \in U$, this equation possess a (true) solution $y = \varphi(x)$ defined for x sufficiently close to x_0, with the initial value

$$\varphi(x_0) = y_0.$$

Briefly, we say: we require that equation (6.5.1) possess a local solution for *arbitrary initial* $(x_0, y_0) \in U$.

FUNDAMENTAL THEOREM 6.5.1. For this to be the case, it is necessary and sufficient that, *for every point* $(x, y) \in U$, the element $f'_x(x, y) + f'_y(x, y) \circ f(x, y) \in \mathscr{L}(E; \mathscr{L}(E; F))$ defines a *symmetric* element of $\mathscr{L}_2(E; F)$; i.e., that

$$(6.5.2)\qquad ((f'_x(x,y) + f'_y(x,y) \circ f(x,y)) \cdot \xi) \cdot \eta = ((f'_x(x,y) + f'_y(x,y) \circ f(x,y)) \cdot \eta) \cdot \xi$$

for arbitrary $\xi, \eta \in E$, and arbitrary $(x, y) \in U$.

This is an obvious consequence of Theorem 6.3.1.

DEFINITION. When condition (6.5.2) is satisfied, we say that equation (6.5.1) is *completely integrable.*

Remark. If E is of dimension 1, the equation is always completely integrable (for then ξ and η are necessarily proportional); the equation (6.5.1) is then an ordinary differential equation.

Complement to the fundamental theorem (without proof). Fix x_0, and allow y to vary: let

$$(6.5.3)\qquad y = \varphi(x, y_0)$$

be the solution equal to y_0 for $x = x_0$. It can be shown that $\varphi(x, y_0)$ is of class C^1, and that $(\partial\varphi/\partial y_0)(x, y_0) \in \text{Isom}(F, F)$, so that (by the Theorem of Implicit Functions) (6.5.3) may be written in the form

$$(6.5.4)\qquad y_0 = \psi(x, y),$$

where ψ is of class C^1. Thus the function $\psi(x, y)$, taking values in F, is *constant* on every solution $y = \varphi(x)$ of equation (6.5.1), supposed completely integrable.

Example. Suppose that $E = \mathbf{R}^p$, $F = \mathbf{R}^{n-p}$, U being an open subset of $\mathbf{R}^n = \mathbf{R}^p \times \mathbf{R}^{n-p}$. The function

$$\psi(x_1, \ldots, x_p, y_1, \ldots, y_{n-p})$$

having values in $\mathbf{R}^{n-p}$, has $(n - p)$ scalar components $\psi_j(x_1, \ldots, x_p, y_1, \ldots, y_{n-p})$, which are constant on all the solutions

$$y_j = \varphi_j(x_1, \ldots, x_p) \qquad (1 \leqslant j \leqslant n - p).$$

These are "first integrals".

6.6 *Interpretation in terms of differential forms*

Again, let us adhere to the notation of § 6.1. There we introduced the differential form

$$\omega = dy - f(x, y) \cdot dx;$$

this is a differential form of degree one, defined in the open $U \subset E \times F$, and with values in F. Its exterior differential is

$$d\omega = -df \underset{\Phi}{\wedge} dx,$$

Φ denoting the bilinear canonical mapping

$$\mathscr{L}(E; F) \times E \to F,$$

which associates to (f, u) the value of $f \in \mathscr{L}(E; F)$ on the vector $u \in E$. Explicitly,

$$d\omega = -\left(\frac{\partial f}{\partial x} dx\right) \underset{\Phi}{\wedge} dx - \left(\frac{\partial f}{\partial y} dy\right) \underset{\Phi}{\wedge} dx.$$

In this replace dy by $f(x, y) \cdot dx$ to obtain the differential form of degree 2:

$$\Omega = -\left(\left(\frac{\partial f}{\partial x} + \frac{\partial f}{\partial y} \circ f\right) \cdot dx\right) \underset{\Phi}{\wedge} dx,$$

with values in F. The value of Ω for a pair $(\xi, \eta) \in E \times E$ (cf. § 2.2) is:

$$-\left(\left(\frac{\partial f}{\partial x} + \frac{\partial f}{\partial y} \circ f\right) \cdot \xi\right) \cdot \eta + \left(\left(\frac{\partial f}{\partial x} + \frac{\partial f}{\partial y} \circ f\right) \cdot \eta\right) \cdot \xi.$$

The condition of complete integrability (6.2.2) expresses the fact that this vanishes. Hence we have:

THEOREM 6.6.1. *In order that the differential equation*

$$\frac{dy}{dx} = f(x, y)$$

be completely integrable, it is necessary and sufficient that the differential 2-*form* Ω (obtained by replacing dy by $f(x, y) \cdot dx$ in $d\omega$) *be identically zero.*

We shall interpret this condition in the case where E and F are of finite dimension. Take $E = \mathbf{R}^p$, and let $x_1, \ldots, x_p$ denote the coordinates of a point in E; take further $F = \mathbf{R}^{n-p}$ and let $x_{p+1}, \ldots, x_n$ be the coordinates of a point in F. Then $x_1, \ldots, x_n$ are the coordinates of a point in $E \times F = \mathbf{R}^n$. Consider an open $U \subset \mathbf{R}^n$; the mapping $f\colon U \to \mathscr{L}(E; F)$ is defined by a matrix $\{f_{ij}(x)\}$ of functions $f_{ij}(x_1, \ldots, x_n)$ of class C^1 in U, and the differential form ω (with values in F) is defined by $n - p$ scalar valued differential forms

$$\boxed{\omega_i = dx_i - \sum_{j=1}^{p} f_{ij}(x)\, dx_j, \qquad p + 1 \leqslant i \leqslant n}. \tag{6.6.1}$$

A solution $y = \varphi(x)$ is defined by $n - p$ functions of class C^1

$$x_i = \varphi_i(x_1, \ldots, x_p) \qquad (p + 1 \leqslant i \leqslant n) \tag{6.6.2}$$

such that the change of variable defined by (6.6.2) annihilates the $n - p$ forms ω_i. In other words, the φ_i must satisfy

$$\frac{\partial \varphi_i}{\partial x_j} = f_{ij}(x_1, \ldots, x_p, \varphi_{p+1}(x_1, \ldots, x_p), \ldots, \varphi_n(x_1, \ldots, x_p)) \tag{6.6.3}$$

$(1 \leqslant j \leqslant p, p + 1 \leqslant i \leqslant n)$. Such a system is *completely integrable* if and only if the differential 2-forms $d\omega_i$ vanish when dx_i is replaced by

$$\sum_{j=1}^{p} f_{ij}(x)\, dx_j,$$

for $i = p + 1, \ldots, n$; in fact, this confirms Theorem 6.6.1. We shall now give an explicit interpretation of this condition.

Let us adjoin to the forms $\omega_{p+1}, \ldots, \omega_n$ a sequence of p forms $\omega_1, \ldots, \omega_p$ in such a manner that

$$\omega_1, \ldots, \omega_p, \omega_{p+1}, \ldots, \omega_n$$

form a *base* for the differential 1-forms (i.e., every differential 1-form may then be written uniquely in the form $\sum_{i=1}^{p} a_i(x)\omega_i$, the coefficients $a_i(x)$ being functions of x): for example, we can take $\omega_1 = dx_1, \ldots, \omega_p = dx_p$ (the reader may verify that $dx_1, \ldots, dx_p$, $\omega_{p+1}, \ldots, \omega_n$ *do* indeed form a base for the differential 1-forms). Then every differential 2-form Ω may be written uniquely in the form

$$\Omega = \sum_{1 \leqslant i < j \leqslant n} b_{ij}(x)\omega_i \wedge \omega_j,$$

where the coefficients b_{ij} are functions of $x_1, \ldots, x_n$ (this is proved as in Theorem 2.6.2, with the aid of Theorem 1.7.1). In order that Ω should vanish when dx_i is replaced by $\sum_{j=1}^{p} f_{ij}(x)\, dx_j$ (for $i = p + 1, \ldots, n$), i.e., when ω_i is replaced by 0 for $p + 1 \leqslant i \leqslant n$, it is necessary and sufficient that the coefficients $b_{ij}(x)$ be identically zero when i, j are both $\leqslant p$. Let us apply this result to $d\omega_i (p + 1 \leqslant i \leqslant n)$; *a priori*, we have

$$d\omega_i = \sum_{1 \leqslant j < k \leqslant n} c_{ijk}(x)\omega_j \wedge \omega_k,$$

and therefore, *the condition of complete integrability is expressed by:*

$$\boxed{c_{ijk}(x) = 0 \quad \text{for} \quad i > p, j \leqslant p, k \leqslant p}.$$

We shall now obtain an elegant expression of these conditions: if we calculate the exterior product $(d\omega_i) \wedge \omega_{p+1} \wedge \cdots \wedge \omega_n$ (for $i > p$), we find

$$(d\omega_i) \wedge \omega_{p+1} \wedge \cdots \wedge \omega_n = \sum_{1 \leqslant j < k \leqslant p} c_{ijk}(x)\omega_j \wedge \omega_{p+1} \wedge \cdots \wedge \omega_n,$$

for if at least one of the integers j and k is $> p$, the exterior product

$$\omega_j \wedge \omega_k \wedge \omega_{p+1} \wedge \cdots \wedge \omega_n$$

vanishes (for it then contains two equal factors). The condition of complete integrability is therefore expressed by the conditions

$$\boxed{(d\omega_i) \wedge \omega_{p+1} \wedge \cdots \wedge \omega_n = 0 \quad \text{for} \quad p + 1 \leqslant i \leqslant n}. \tag{6.6.4}$$

In fact, every $(n - p + 2)$-form may be written uniquely as a linear combination of

$$\omega_{k_1} \wedge \omega_{k_2} \wedge \cdots \wedge \omega_{k_{n-p+2}} \qquad (1 \leqslant k_1 < \cdots < k_{n-p+2} \leqslant n),$$

the coefficients being functions of x; the necessary and sufficient condition that such a form be zero is that all its coefficients vanish.

In the condition (6.6.4), recall that the ω_i denote the differential forms

$$dx_i - \sum_{j=1}^{p} f_{ij}(x)\, dx_j \qquad (p + 1 \leqslant i \leqslant n);$$

suppose now that we make a linear transformation of the $\omega_{p+1}, \ldots, \omega_n$,

$$\alpha_i = \sum_{j=p+1}^{n} u_{ij}(x)\omega_j,$$

the matrix $u_{ij}(x)$ having non-zero determinant for all x. The system $\omega_i = 0$ $(p + 1 \leqslant i \leqslant n)$ is equivalent to the system $\alpha_i = 0$ $(p + 1 \leqslant i \leqslant n)$. We claim that *the condition of complete integrability is equivalent to*

$$(6.6.5) \qquad d\alpha_i \wedge \alpha_{p+1} \wedge \cdots \wedge \alpha_n = 0 \quad \text{for} \quad p + 1 \leqslant i \leqslant n.$$

Indeed,

$$\alpha_{p+1} \wedge \cdots \wedge \alpha_n = \det(u_{ij}(x))\omega_{p+1} \wedge \cdots \wedge \omega_n,$$

therefore (6.6.5) is equivalent to

$$d\alpha_i \wedge \omega_{p+1} \wedge \cdots \wedge \omega_n = 0.$$

Further

$$d\alpha_i = \sum_j u_{ij}\, d\omega_j + \sum_j (du_{ij}) \wedge \omega_j,$$

therefore,

$$d\alpha_i \wedge \omega_{p+1} \wedge \cdots \wedge \omega_n = \sum_{j>p} u_{ij} \cdot d\omega_j \wedge \omega_{p+1} \wedge \cdots \wedge \omega_n.$$

Thus the system (6.6.4) implies (6.6.5): the same reasoning establishes the converse.

Q.E.D.

In summary (and writing now ω_i in place of α_i):

THEOREM 6.6.2. *Suppose* $(\omega_{p+1}, \ldots, \omega_n)$ *is a system of* $n - p$ *differential forms of degree* 1, *of class* C^1, *in an open* $U \subset \mathbf{R}^n$, *such that, at each point* $x \in U$, *the rank of the system* $(\omega_{p+1}, \ldots, \omega_n)$ *is equal to* $n - p$. *Then the differential system*

$$\omega_i = 0 \qquad (p + 1 \leqslant i \leqslant n)$$

(system of "total differential equations") *is completely integrable if, and only if, the differential forms*

$$d\omega_i \wedge \omega_{p+1} \wedge \cdots \wedge \omega_n \qquad (p + 1 \leqslant i \leqslant n)$$

vanish identically ("the condition of Frobenius").

Example. Suppose ω is a differential form of degree 1 and of class C^1:

$$\omega = \sum_{i=1}^{n} a_i(x)\, dx_i,$$

whose coefficients $a_i(x)$ do not all vanish simultaneously. *In order that the equation* $\omega = 0$ *be completely integrable, it is necessary and sufficient that*

$$\omega \wedge d\omega = 0.$$

(*Exercise.* Show that this is also the condition for ω to possess an "integrating factor", i.e., in order that there exist a function $\mu(x) \neq 0$ such that the 1-form $\mu(x)\cdot\omega$ is closed.)

More particularly, for $n = 3$ (and denoting the coordinates of a point in $\mathbf{R}^3$ by x, y, z), let

$$\omega = P(x, y, z)\, dx + Q(x, y, z)\, dy + R(x, y, z)\, dz.$$

In order that the equation $\omega = 0$ be completely integrable, it is necessary and sufficient that

$$(6.6.6) \qquad P\left(\frac{\partial R}{\partial y} - \frac{\partial Q}{\partial z}\right) + Q\left(\frac{\partial P}{\partial z} - \frac{\partial R}{\partial x}\right) + R\left(\frac{\partial Q}{\partial x} - \frac{\partial P}{\partial y}\right) = 0.$$

(The vector (P, Q, R) must be orthogonal to its "curl".) Indeed, it is immediately clear that the left-hand side of (6.6.6) is the coefficient of $dx \wedge dy \wedge dz$ in the canonical expression of the form $\omega \wedge d\omega$.

Exercises

1. Let $\alpha_1, \ldots, \alpha_p$, $p \leqslant n$ be p linear independent forms on $\mathbf{R}^n$. Show that for a linear form α on $\mathbf{R}^n$ it is true that

$$\alpha \wedge \alpha_1 \wedge \cdots \wedge \alpha_p = 0$$

if and only if α belongs to the vector space generated by $\alpha_1, \ldots, \alpha_p$.

Show also that, if this condition is satisfied and if $\alpha \neq 0$, there exists a linear alternating $(p - 1)$-form β such that

$$\alpha_1 \wedge \cdots \wedge \alpha_p = \alpha \wedge \beta.$$

2. Consider the differential 2-form in $\mathbf{R}^n$

$$\omega = dx_1 \wedge dx_2 + dx_3 + dx_4 + \cdots + dx_{2n-1} \wedge dx_{2n}.$$

Calculate $\overset{n}{\wedge}\, \omega$, the outer product of n factors equal to ω.

3. Let $\alpha_1, \ldots, \alpha_n$ be n linear independent forms on $\mathbf{R}^n$; consider the bilinear alternating form

$$\omega = \sum_{j=1}^{n} \sum_{i=1}^{j-1} a_{ij}\alpha_i \wedge \alpha_j,$$

where the a_{ij}, $i < j$, are $\binom{n}{2}$ real given numbers.

(a) Show that if $a_{12} \neq 0$ there exist two linear forms β_1 and β_2 such that the n linear forms $\beta_1, \beta_2, \alpha_3, \ldots, \alpha_n$ are also linearly independent and $\omega - \beta_1 \wedge \beta_2$ may be expressed in terms of $\alpha_3, \ldots, \alpha_n$ only.

Deduce by induction that there exist $2r$ linearly independent forms, with $r \leqslant \frac{1}{2}n$ if n is even and $r \leqslant \frac{1}{2}(n - 1)$ if n is odd, such that

$$\omega = \beta_1 \wedge \beta_2 + \beta_3 \wedge \beta_4 + \cdots + \beta_{2r-1} \wedge \beta_{2r}.$$

(b) Introduce the antisymmetric matrix $A = (A_{ij})$ defined by $A_{ij} = a_{ij}$ for $i < j$. Show that the number $2r$ defined in (a) is equal to the rank of the matrix A and is independent of the choice of the forms comprising the base $\alpha_1, \ldots, \alpha_n$.

(c) Show that the number r considered above is equal to the smallest integer r such that

$$\overset{r+1}{\wedge}\ \omega = 0.$$

Deduce that $\omega \wedge \omega = 0$ if and only if ω is equal to the exterior product of two linear forms.

4. Let $\omega \neq 0$ be a linear form on $\mathbf{R}^n$ and let α be a q-linear alternating form on $\mathbf{R}^n$. Show that the condition $\omega \wedge \alpha = 0$ is necessary and sufficient in order for there to exist a linear alternating $(q-1)$-form β such that

$$\alpha = \omega \wedge \beta.$$

5. Let α be the differential 1-form in $\mathbf{R}^n$ defined by

$$\alpha = y\,dx - x\,dy + dz.$$

(a) Under what conditions (C) do the functions $u(x, y, z)$ and $v(x, y, z)$ of class C^1 make the form

$$\alpha - v\,du$$

closed? Show that u and v are then independent of z.
(b) Is it possible for $v = V(x, y)$ to be chosen arbitrarily?
(c) Show that if u and v satisfy conditions (C) the three differential forms du, dv and $\alpha - v\,du$ are linearly independent at each point.

6. Let

$$\omega = a\,dy \wedge dz + b\,dz \wedge dx + c\,dx \wedge dy$$

be a differential form of class C^∞ on $\mathbf{R}^3$; let M_0 denote a point of $\mathbf{R}^3$ where ω is non-zero. Let f be a function of class C^∞ defined in a neighbourhood of M_0.
(a) Show that, in order that ω may be expressed in the form $\alpha \wedge df$ in the neighbourhood of M_0, α being a differential 1-form of class C^∞ in the neighbourhood of M_0, it is necessary and sufficient that f does not vanish at M_0, and that f is the solution of a certain partial differential equation. Write down this equation.
(b) Set $\alpha = \lambda\,dx + \mu\,dy + \nu\,dz$; if $\alpha \wedge df = \omega$, express λ, μ, ν as functions of $a, b, c, \partial f/\partial x, \partial f/\partial y, \partial f/\partial z$.

7. Let f be a real function of class C^2 in an open neighbourhood Ω of a point $x^{(0)}$ of $\mathbf{R}^n$. Put $u_i(x) = (\partial f/\partial x_i)(x)$ and let φ be the mapping $x \mapsto u = (u_1, \ldots, u_n)$.

Under what condition does there exist a neighbourhood V of $x^{(0)}$ such that φ is a diffeomorphism of V onto $\varphi(V)$?

Suppose this condition is satisfied and for every $u \in \varphi(v)$ put $x = \varphi^{-1}(u)$. Show that the differential form

$$\omega = \sum_{i=1}^{n} x_i\,du_i$$

is closed. Deduce that, in a neighbourhood V of $u^{(0)} = \varphi(x^{(0)})$, there exists a function g of class C^2 such that $x_i = \partial g/\partial u_i$.

Show that if f is a homogeneous function of degree $p \neq 1$, then, in $\varphi(U)$

$$g \circ \varphi = (p - 1)f + \text{const.}$$

and that g can be chosen to be homogeneous of degree $p/(p-1)$.

Determine $g(u)$ for $f = x_1x_2x_3$.

8. Let U be an open subset of $\mathbf{R}^n$ which is starlike with respect to the origin.
(a) Suppose $n = 3$ and let

$$\omega = A\,dy \wedge dz + B\,dz \wedge dx + C\,dx \wedge dy$$

be a differential 2-form of class C^q, $q \geqslant 1$, in V. Apply the operator k of Prop. 2.13.2 to obtain the functions P, Q, R such that

$$k(\omega) = P\,dx + Q\,dy + R\,dz.$$

(b) Now let ω be the differential p-form of class C^q in V:

$$\omega = c(x)\,dx_1 \wedge \cdots \wedge dx_p.$$

Show that the differential $(p-1)$-form $k(\omega)$ has the canonical form

$$k(\omega) = a(x) \sum_{i=1}^{p} (-1)^{i-1} x_i\,dx_1 \wedge \cdots \wedge dx_{i-1} \wedge dx_{i+1} \wedge \cdots \wedge dx_p,$$

where

$$a(x) = \int_0^1 t^{p-1} c(tx)\,dt.$$

9. In $\mathbf{R}^n - \{0\}$, consider the differential $(n-1)$-form

$$\omega = \sum_{i=1}^{n} (-1)^{i-1} \frac{x_i}{r^\alpha}\,dx_1 \wedge \cdots \wedge dx_{i-1} \wedge dx_{i+1} \wedge \cdots \wedge dx_n, \qquad r = \sqrt{x_1^2 + \cdots + x_n^2}.$$

(a) How must α be chosen in order that $d\omega = 0$? Having made this choice express the integral of ω on the unit sphere of $\mathbf{R}^n$ (with the usual orientation) as an integral in $dx_1 \wedge \cdots \wedge dx_n$.

Calculate the preceding integral for $n = 2$, $n = 3$.

(b) Suppose that $n = 3$ and that α has the form determined in the previous question. Apply the formula established in the previous exercise to obtain a primitive of ω in an open set starlike with respect to the point (0, 0, 1).

10. Let U be an open set in $\mathbf{R}^n$ which is starlike with respect to the origin and let $X = (X_1, \ldots, X_n)$ be a vector field of class C^1 in U.

To every p-form

$$\omega = \sum a_{i_1 \ldots i_p}\,dx_{i_1} \wedge \cdots \wedge dx_{i_p},$$

of class C^1 in U make the correspondence

$$i(X)\cdot\omega = \sum a_{i_1 \ldots i_p} \sum_{k=1}^{p} (-1)^{k-1} X_{i_k}\,dx_{i_1} \wedge \cdots \wedge \widehat{dx_{i_k}} \wedge \cdots \wedge dx_{i_p}, \quad \text{if} \quad p \geqslant 1,$$

$$i(X)\cdot\omega = 0 \quad \text{if} \quad p = 0.$$

(a) Show that $i(X)\cdot i(X) = 0$ and that

$$i(X)\cdot(\alpha \wedge \beta) = (i(X)\cdot\alpha) \wedge \beta + (-1)^p \alpha \wedge i(X)\cdot\beta,$$

for every p-form α.

Put

$$X\cdot\omega = i(X)\cdot d\omega + d(i(X)\cdot\omega).$$

Show that

$$X \circ i(X) = i(X) \circ X$$
$$X \circ d = d \circ X$$
$$X\cdot(\alpha \wedge \beta) = (X\cdot\alpha) \wedge \beta + \alpha \wedge (X\cdot\beta).$$

(b) In the following we shall be particularly interested in the operator X_0 associated with the vector field $X_0(x) = x$.

Show that

$$X_0\cdot(a\,dx_1 \wedge \cdots \wedge dx_p) = \left(pa + \sum_{i=1}^{n} x_i \frac{\partial a}{\partial x_i}\right) dx_1 \wedge \cdots \wedge dx_p.$$

Show that the partial differential equation

$$pf + \sum_{i=1}^{n} x_i \frac{\partial f}{\partial x_i} = h,$$

h being a given function, possesses a unique solution of class C^1 in U, equal to

$$h^*(x) = \int_0^1 h(tx_1, \ldots, tx_n)t^{p-1}\,dt.$$

Deduce that, for a given form ω_1 of degree $\geqslant 1$ in U, there exists a unique form ω such that $X_0\cdot\omega = \omega_1$; put $\omega = X_0^{-1}\cdot\omega_1$. Show that

$$X_0^{-1} \circ i(X_0) = i(X_0) \circ X_0^{-1}$$
$$d \circ X_0^{-1} = X_0^{-1} \circ d.$$

(c) Put

$$k(\omega) = (i(X_0) \circ X_0^{-1})\cdot\omega \quad \text{for} \quad p \geqslant 1,$$
$$k(\omega) = 0 \quad \text{for} \quad p = 1.$$

Show that

$$d(k(\omega)) + k(d\omega) = \omega, \quad \text{for} \quad p \geqslant 1,$$

and verify that k is the operator of Prop. 2.13.2. Thus we have another proof of Poincaré's theorem.

11. Calculate the area and volume of the torus defined by the parametric equations

$$\begin{cases} x = (a + R\cos\theta)\cos\varphi \\ y = (a + R\cos\theta)\sin\varphi \\ z = R\sin\theta, \end{cases}$$

for $R < a$.

12. In $\mathbf{R}^3$, formulate and prove Stokes's theorem for a cylinder whose generators are parallel to $0z$ and whose base is the boundary ∂K of the compact set with boundary K of $\mathbf{R}^2$.

13. Let $g = (g_1, \ldots, g_n)$ be a mapping of class C^2 of a neighbourhood of the closed unit ball of $\mathbf{R}^n$ into the unit sphere S.

Calculate the integrals

$$\int_S g_1\,dg_2 \wedge \cdots \wedge dg_n \quad \text{and} \quad \int_S x_1\,dx_2 \wedge \cdots \wedge dx_n.$$

(S has the usual orientation.) Deduce that there does not exist such a mapping g such that its relation to S is equal to the identity mapping.

14. Let P, Q, R denote three functions of class C^1 in an open set U of $\mathbf{R}^3$, which do not vanish simultaneously; put

$$\omega = P\,dy \wedge dz + Q\,dz \wedge dx + R\,dx \wedge dy.$$

(a) Show that in the neighbourhood of each point of U there exist pairs of functions u, v satisfying

$$du \wedge \omega = 0 \qquad dv \wedge \omega = 0 \qquad du \wedge dv \neq 0,$$

and show that if u, v is such a pair, there exists a function λ such that

$$\omega = \lambda\,du \wedge dv.$$

Deduce that in the neighbourhood of each point U there exist two functions g, h such that $\omega = dg \wedge dh$ if and only if $d\omega = 0$.
(b) Let f be a function of class C^1 in U whose partial derivatives f'_x, f'_y, f'_z do not vanish simultaneously; and let V_a be the variety (surface) defined implicitly by $f(x, y, z) = a$.
Show that

$$df \wedge \omega = 0 \quad \text{implies that} \quad \iint_{V_a \cap \Omega} \omega = 0$$

for arbitrary number $a \in f(U)$, and sufficiently small open Ω. Is the converse also true?

15. If a, b, c are given constants, determine all the linear transformations of $\mathbf{R}^3$ which preserve the differential form

$$\omega = a\, dy \wedge dz + b\, dz \wedge dx + c\, dx \wedge dy.$$

16. In $\mathbf{R}^3$ consider the differential form

$$\omega = x\, dy \wedge dz - 2zf(y)\, dx \wedge dy + yf(y)\, dz \wedge dx,$$

where f is a mapping of class C^1 of $\mathbf{R}$ into $\mathbf{R}$ such that $f(1) = 1$.
(a) Determine f if $d\omega = dx \wedge dy \wedge dz$. For this choice of f, calculate the integral $\int_S \omega$, where S denotes the spherical cap

$$x^2 + y^2 + z^2 = 1, \qquad z \geqslant \sqrt{2}/2,$$

oriented so that the normal is directed away from the centre of the sphere.
(b) Determine f if $d\omega = 0$ and, for this choice of f, repeat the calculation of $\int_S \omega$.
(c) Determine f such that there exists a form $\omega_1 = P(x, y, z)\, dx + Q(x, y, z)\, dy$, with

$$P(x, y, 0) = Q(x, y, 0) = 0$$

and $d\omega_1 = \omega$. Then calculate $\int_C \omega_1$, where C is the circumference

$$x^2 + y^2 + z^2 = 1, \qquad z = \sqrt{2}/2$$

oriented in such a manner that it is the oriented boundary of S.

17. Consider the differential form of class C^3, given in a neighbourhood V of the origin of $\mathbf{R}^3$:

$$\omega = P(x, y, z)\, dx + Q(x, y, z)\, dy + R(x, y, z)\, dz.$$

Suppose that

$$\frac{\partial R}{\partial y} - \frac{\partial Q}{\partial x} \neq 0.$$

(a) In the neighbourhood of 0, obtain the surfaces

$$(S) \quad z = f(x, y)$$

of class C^2, such that the differential form induced on S by ω is *closed* [the induced form is that obtained from ω by making the change of variable $(x, y, z) \mapsto (x, y, f(x, y))$]. Show that f is the solution of a partial differential equation, and obtain the system of characteristics (Γ).
(b) Show that, in a sufficiently small neighbourhood of the origin, there exists a C^2-diffeomorphism mapping the integral curves of (Γ) into the straight lines y = const., z = const. In the following suppose that the characteristic curves are straight lines y = const., z = const.,

and that ω is of class C^2; what are the surfaces S? Show that in the neighbourhood of 0, there exists a function $T(y, z)$ of class C^1, vanishing at the origin, such that

$$d\omega = \frac{\partial T}{\partial y}\, dy \wedge dz, \quad \text{with} \quad \frac{\partial T}{\partial y}(0, 0) \neq 0.$$

Deduce that there exists a function $U(x, y, z)$ of class C^1, vanishing at the origin, such that

$$\omega = dU + T(y, z)\, dz. \tag{1}$$

(c) In the following suppose that ω is given by equation (1), where T and U satisfy the conditions stated above. Show that if $\partial U/\partial x \neq 0$, the equation $\omega = 0$ is not completely integrable.

Under the hypothesis $\partial U/\partial x \neq 0$, show that there exists a C^1-diffeomorphism of a neighbourhood of the origin:

$$x = \lambda(X, Y, Z), \qquad y = \mu(Y, Z), \quad z = Z$$

such that, by this change of variable, ω becomes $dX + Y\, dZ$.

18. Let $U \subset \mathbf{R}^3$ be the open set formed by the points (x, y, z) such that $xyz \neq 0$; in U, consider the form

$$\omega = \frac{1}{yz}\, dx + \frac{1}{zx}\, dy + \frac{1}{xy}\, dz.$$

(a) Show that the equation $\omega = 0$ is completely integrable.
(b) Determine an integrating factor for ω, i.e., a function f of class C^1 in U such that $d(f\omega) = 0$ (in the partial differential equation for f, the substitution $\varphi = \log f$ can be made).

19. Under what conditions (2) is the system

$$\left.\begin{aligned} dz_1 &= a\, dx_1 + b\, dx_2 \\ dz_2 &= a\, dy_1 + b\, dy_2 \end{aligned}\right\} \tag{1}$$

completely integrable, a and b being functions of class C^1 of the variables $x_1, x_2, y_1, y_2, z_1, z_2$?

Integrate the system under conditions (2) (express z_1, z_2, as functions of a, b, x_1, x_2, y_1, y_2, and show that they are linear functions of x_1, x_2, y_1, y_2).

20. Let P and Q be two functions of class C^1 of the four variables x, y, u, v. Under what conditions is the system

$$\begin{cases} dx = P\, du - Q\, dv \\ dy = Q\, du + P\, dv \end{cases}$$

completely integrable?

Set $z = x + iy$, $\omega = u + iv$, $f(z, \omega) = P + iQ$. Show that, in particular, the preceding conditions are satisfied if $f(z, \omega)$ is a regular function of z and ω.

21. *Preliminary remark:* a vector field in an open Ω of a Banach space E is a mapping of Ω into E; if it exists, its derivative at a point is therefore a linear mapping of E into E.

In the following, Ω is an open ball centre x_0 and of radius r.

(a) Let U be a vector field of class C^1 in Ω; consider the solution $\varphi(t, x_0)$ of the equation $dx/dt = U(x)$ equal to x_0 for $t = 0$.

Show that φ has the second order expansion near $t = 0$

$$\varphi(t, x_0) = x_0 + tU(x_0) + \frac{t^2}{2} U'(x_0) \cdot U(x_0) + o(t^2).$$

(b) Let U and V be two vector fields of class C^1 in Ω such that $\|U(x)\| \leqslant M$ and $\|V(x)\| \leqslant M$ for $x \in \Omega$. For

$$\theta \in \left]\frac{-r}{4M}, \frac{+r}{4M}\right[,$$

let x_1 be the value for $t = \theta$ of the solution of $dx/dt = U(x)$ which is equal to x_0 for $t = 0$, x_2 the value for $t = \theta$ of the solution of $dx/dt = V(x)$ which is equal to x_1 for $t = 0$, x_3 the value for $t = \theta$ of the solution of $dx/dt = -U(x)$ which is equal to x_2 for $t = 0$, x_4 the value for $t = \theta$ of the solution of $dx/dt = -V(x)$ which is equal to x_3 for $t = 0$.

Using the expansions of the x_i, show that

$$x_4 = x_0 + \theta^2(V'(x_0)\cdot U(x_0) - U'(x_0)\cdot V(x_0)) + o(\theta^2).$$

What is the tangent at the point x_0 to the curve described by x_4 when θ varies in the interval indicated above?

(c) Suppose that

$$[U, V](x) = V'(x)\cdot U(x) - U'(x)\cdot V(x) = 0$$

for all $x \in \Omega$. By applying the Theorem of Frobenius, show that there exists a function $F(u, v)$ of class C^2 in a neighbourhood of the origin with values in Ω such that

$$F(0, 0) = x_0, \qquad \frac{\partial F}{\partial u}(u, v) = U(F(u, v)), \qquad \frac{\partial F}{\partial v}(u, v) = V(F(u, v)).$$

Write x_1, x_2, x_3, x_4 in terms of F and show that $x_4 = x_0$.

Chapter 2

Elements of the calculus of variations

1 Introduction to the problem

1.1 *The space of curves of class* C^1

Let $I = [a, b] \subset \mathbf{R}$ be a compact interval. A curve (parametrized by $t \in I$) in the space $\mathbf{R}^n$ is a mapping

$$\varphi: I \to \mathbf{R}^n;$$

it is of class C^k if φ is of class C^k. More generally, let us consider curves in a Banach space E (on $\mathbf{R}$): by definition, a curve of class C^1 is a mapping

$$\varphi: I \to E$$

of class C^1. It is evident that the set of these curves form a *vector space* (on $\mathbf{R}$). We shall denote this vector space by V (we shall omit reference to E in this notation, which we shall suppose given once and for all).

We shall proceed to construct a *Banach space* on V. To do this we must define a *norm* on V, and show that V is *complete* with respect to this norm.

DEFINITION. For $\varphi: I \to E$ of class C^1, put

$$\|\varphi\| = \sup_{t \in I} \|\varphi(t)\| + \sup_{t \in I} \|\varphi'(t)\|; \tag{1.1.1}$$

this is a *finite* real number $\geqslant 0$, since $t \mapsto \|\varphi(t)\|$ and $t \mapsto \|\varphi'(t)\|$ are continuous non-negative functions, which are therefore bounded on the compact interval I. The reader may verify that $\|\varphi\|$ is indeed a *norm* on V.

PROPOSITION 1.1.1. V *is complete with respect to the norm defined by* (1.1.1).

PROOF. Let (φ_m) be a Cauchy sequence; since

$$\sup_{t \in I} \|\varphi_m(t) - \varphi_n(t)\| \leqslant \|\varphi_m - \varphi_n\|,$$

the sequence (φ_n) is also a Cauchy sequence for the norm of the uniform convergence; therefore φ_n converges uniformly to a *continuous* function $\varphi: I \to E$ (since E is complete). It remains to show that φ is of class C^1, and is the limit of the sequence (φ_n) for the norm of V.

Now the sequence of derivatives is also a Cauchy sequence for the norm of uniform convergence, since

$$\sup_{t \in I} \|\varphi'_m(t) - \varphi'_n(t)\| \leqslant \|\varphi_m - \varphi_n\|.$$

Therefore φ'_n converges uniformly to a function ψ which is continuous. By Theorem 3.6.1 of Chap. 1 of *Differential Calculus*, φ possesses a derivative φ' which is equal to ψ. Thus,

$$\lim_{n \to \infty} \sup_{t \in I} \|\varphi_n(t) - \varphi(t)\| = 0,$$

$$\lim_{n \to \infty} \sup_{t \in I} \|\varphi'_n(t) - \varphi'(t)\| = 0,$$

and it follows that

$$\lim_{n \to \infty} \|\varphi_n - \varphi\| = 0,$$

which completes the proof.

1.2 *Functional associated with a curve*

In the following we shall suppose that in an open $U \subset \mathbf{R} \times E \times E$ we are given a function

$$F: U \to \mathbf{R}$$

of class C^k. We shall denote by $F(t, x, y)$ the value of this function at a point $(t, x, y) \in U$.

Given a curve $\varphi: I \to E$ of class C^1 such that

$$(1.2.1) \qquad (t, \varphi(t), \varphi'(t)) \in U \quad \text{for} \quad t \in I,$$

we can associate it with a real number

$$\int_a^b F(t, \varphi(t), \varphi'(t))\, dt.$$

This number depends on the choice of φ; denote it by $f(\varphi)$:

$$(1.2.2) \qquad f(\varphi) = \int_a^b F(t, \varphi(t), \varphi'(t))\, dt.$$

If Ω denotes the set of $\varphi \in V$ which satisfies (1.2.1), then, by means of F, we have thus defined a mapping

$$f: \Omega \to \mathbf{R}.$$

This is a function defined on the set of functions Ω or (as we sometimes say) a "functional" defined on Ω. We propose to study f in more detail.

PROPOSITION 1.2.1. Ω is an *open* set of the Banach space V.

PROOF. Let $\varphi_0 \in \Omega$; it is required to show that every $\varphi \in V$ such that $\|\varphi - \varphi_0\|$ is sufficiently small also belongs to Ω. Now let $K \subset U$ be the image of the continuous mapping

$$t \to (t, \varphi(t), \varphi'(t))$$

of I into Ω; K is compact, since I is compact. Therefore there exists $\rho > 0$ such that every point $(t, x, y) \in \mathrm{I} \times \mathrm{E} \times \mathrm{E}$ satisfying

$$\|x - \varphi_0(t)\| \leqslant \rho, \qquad \|y - \varphi_0'(t)\| \leqslant \rho$$

belongs to U (a direct proof may be given using the compactness of I). Now suppose that $\varphi \in \mathrm{V}$ satisfies $\|\varphi - \varphi_0\| \leqslant \rho$; then

$$\|\varphi(t) - \varphi_0(t)\|_{\mathrm{E}} \leqslant \rho \quad \text{and} \quad \|\varphi'(t) - \varphi_0'(t)\|_{\mathrm{E}} \leqslant \rho$$

or all $t \in \mathrm{I}$, by the definition of the norm of V. Thus we have indeed

$$(t, \varphi(t), \varphi'(t)) \in \mathrm{U} \quad \text{for all} \quad t \in \mathrm{I}.$$

Q.E.D.

Thus the mapping f is defined in an open set of a Banach space, and it is meaningful to ask if it is of class C^k.

PROPOSITION 1.2.2. *If* $\mathrm{F}: \mathrm{U} \to \mathbf{R}$ *is of class* C^k $(k \geqslant 1)$, *the mapping* $f: \Omega \to \mathbf{R}$ *defined by* (1.2.2) *is also of class* C^k; *further, the derivative* f' *is given by*

$$(1.2.3) \quad f'(\varphi)\cdot u = \int_a^b \frac{\partial \mathrm{F}}{\partial x}(t, \varphi(t), \varphi'(t))\cdot u(t)\, dt + \int_a^b \frac{\partial \mathrm{F}}{\partial y}(t, \varphi(t), \varphi'(t))\cdot u'(t)\, dt$$

where $u \in \mathrm{V}$.

(To understand this formula, recall that, for $\varphi \in \Omega$, $f'(\varphi)$ must be an element of $\mathscr{L}(\mathrm{V}; \mathbf{R})$; if u is an element of V, u is a function of class C^1: $\mathrm{I} \to \mathrm{E}$; u' denotes the derived mapping, and, for $t \in \mathrm{I}$, $u(t)$ and $u'(t)$ are elements of E. As for $(\partial\mathrm{F}/\partial x)(t, \varphi(t), \varphi'(t))$, this is an element of $\mathscr{L}(\mathrm{E}; \mathbf{R})$. The value of this element appears in the right-hand side of (1.2.3) for $u(t) \in \mathrm{E}$. Similarly $(\partial\mathrm{F}/\partial y)(t, \varphi(t), \varphi'(t))$ is an element of $\mathscr{L}(\mathrm{E}; \mathbf{R})$ whose value corresponds to $u'(t) \in \mathrm{E}$.)

PROOF of Prop. 1.2.2. First we shall show that if F is of class C^1, f possesses a derivative given by (1.2.3), which shows that the mapping $\varphi \to f'(\varphi)$ of Ω into $\mathscr{L}(\mathrm{V}; \mathbf{R})$ is *continuous* (*Exercise:* verify this); this shows that f is of class C^1.

In the proof we shall use the lemma for differentiation under the sign of integration (Chap. 1, Lemma 2.12.2). To facilitate this, let us introduce the mapping

$$\lambda: \Omega \times \mathrm{I} \to \mathbf{R}$$

defined by

$$\lambda(\varphi, t) = \mathrm{F}(t, \varphi(t), \varphi'(t)).$$

Then,

$$f(\varphi) = \int_a^b \lambda(\varphi, t)\, dt,$$

the integral of a function λ which depends on a parameter φ varying in an open Ω of a Banach space V. Lemma 2.12.2 says that if the derivative $\partial\lambda/\partial\varphi$ exists and is a continuous function of $(\lambda, t) \in \Omega \times \mathrm{I}$, then f' exists and

$$f'(\varphi) = \int_a^b \frac{\partial \lambda}{\partial \varphi}(\varphi, t)\, dt,$$

or, equivalently,

$$f'(\varphi)\cdot u = \int_a^b \left(\frac{\partial\lambda}{\partial\varphi}(\varphi, t)\cdot u\right) dt \quad \text{for all} \quad u \in \mathrm{V}. \tag{1.2.4}$$

Therefore, let us first establish the existence of $\partial\lambda/\partial\varphi$ by calculation. λ is a composite mapping

$$\Omega \times \mathrm{I} \xrightarrow{\mu} \mathrm{U} \xrightarrow{\mathrm{F}} \mathbf{R},$$

where $\mu(\varphi, t) = (t, \varphi(t), \varphi'(t))$; by hypothesis, F is of class C^1, and it is therefore sufficient to show that $\partial\mu/\partial\varphi$ exists, and to calculate it. Of the three components of $\mu(\varphi, t)$, the first, t, does not depend on φ; the second is $\varphi(t)$, which is a linear continuous function of $\varphi \in \mathrm{V}$, therefore its derivative is constant, and is an element of $\mathscr{L}(\mathrm{V}; \mathbf{R})$ whose value for $u \in \mathrm{V}$ is $u(t)$. Finally the third component is $\varphi'(t)$, which is a linear continuous function of $\varphi \in \mathrm{V}$ (because of the type of norm chosen on V), therefore its derivative is constant, and is an element of $\mathscr{L}(\mathrm{V}; \mathbf{R})$ whose value for $u \in \mathrm{V}$ is $u'(t)$. Thus from the formula for the differential of a function of a function

$$\frac{\partial\lambda}{\partial\varphi}(\varphi, t)\cdot u = \frac{\partial \mathrm{F}}{\partial x}(t, \varphi(t), \varphi'(t))\cdot u(t) + \frac{\partial \mathrm{F}}{\partial y}(t, \varphi(t), \varphi'(t))\cdot u'(t); \tag{1.2.5}$$

this shows that $(\partial\lambda/\partial\varphi)(\varphi, t) \in \mathscr{L}(\mathrm{V}; \mathbf{R})$ is indeed a *continuous* function of $(\varphi, t) \in \Omega \times \mathrm{I}$. Thus relation (1.2.3) follows as a consequence of (1.2.4) and (1.2.5).

The same lemma shows that if $\lambda(\varphi, t)$ is k-times differentiable with respect to φ, and if its derivatives are continuous functions of (φ, t), the integral $\int_a^b \lambda(\varphi, t)\, dt$ is a function of class C^k of the "parameter" φ. This proves Prop. 1.2.2.

1.3 *Example*

Let us now give a very simple example (others will be given later).

Let $\mathrm{E} = \mathbf{R}^n$; take $\mathrm{U} = \mathrm{I} \times \mathbf{R}^n \times (\mathbf{R}^n - \{0\})$ and

$$\mathrm{F}(t, x, y) = \sqrt{(y_1)^2 + \cdots + (y_n)^2}$$

(where $x_1, \ldots, x_n$ are the coordinates of $x \in \mathbf{R}^n$, and $(y_1, \ldots, y_n)$ are the coordinates of $y \in \mathbf{R}^n$). F is of class C^∞ (since the point $y = 0$ has been excluded). The set Ω consists of the curves $\varphi\colon \mathrm{I} \to \mathbf{R}^n$ of class C^1, such that $\varphi'(t) \neq 0$ for all $t \in \mathrm{I}$. We have

$$f(\varphi) = \int_a^b \sqrt{\varphi_1'(t)^2 + \cdots + \varphi_n'(t)^2}\, dt;$$

$f(\varphi)$ is simply the *arc length of the curve* φ. It is a C^∞ function of $\varphi \in \Omega$ (V having the norm given in § 1.1).

1.4 *A problem of minimum*

The set of curves $\varphi\colon \mathrm{I} \to \mathrm{E}$ such that

$$\varphi(a) = \alpha, \qquad \varphi(b) = \beta$$

(where $\alpha \in E$ and $\beta \in E$ are given) is an *affine subspace* $W(\alpha, \beta)$ of the Banach space V. It is a space of codimension 2, since a value α (resp. β) has been imposed on the linear continuous form

$$\varphi \mapsto \varphi(a) \qquad (\text{resp. } \varphi \mapsto \varphi(b)).$$

If a $\varphi_0 \in W(\alpha, \beta)$ is chosen, every element of $W(\alpha, \beta)$ can be set in the form

$$\varphi = \varphi_0 + \psi,$$

with $\psi \in W(0, 0)$, which is a *vector subspace*. Thus φ_0 defines a bijection $W(0, 0)$ onto $W(\alpha, \beta)$; this is the bijection defined by the translation

$$\psi \to \varphi_0 + \psi.$$

$W(\alpha, \beta)$ is none other than the space of curves $\varphi: I \to E$ of class C^1 whose origin (a, α) and extremity (b, β) are given. With respect to the metric defined by the norm of V, $W(\alpha, \beta)$ is a *complete* space. We shall be interested in

$$W(\alpha, \beta) \cap \Omega = \Omega(\alpha, \beta),$$

which is an *open* subset of $W(\alpha, \beta)$. We shall consider the restriction of $f: \Omega \to \mathbf{R}$ to $\Omega(\alpha, \beta)$; if $F: U \to \mathbf{R}$ is of class C^k, then f is of class C^k (Prop. 1.2.2), therefore the restriction of f to the open subset $\Omega(\alpha, \beta)$ of the affine space $W(\alpha, \beta)$ is of class C^k.

(*Remark.* The differentiation of functions in an open subspace of an affine and closed subspace of a Banach space V is well defined, since by translation it becomes an open subset of a closed vector subspace of V, which is itself a Banach space.)

Consider a curve $\varphi_0 \in W(\alpha, \beta)$; we can ask the question: if f is restricted to $W(\alpha, \beta)$ does it possess a relative minimum at φ_0, i.e., is it true that

$$f(\varphi) \geqslant f(\varphi_0)$$

for all $\varphi \in W(\alpha, \beta)$ sufficiently close to φ_0? We propose to give a *necessary condition* for the existence of such a minimum. We know that the restriction $f_{\alpha,\beta}$ of f to $W(\alpha, \beta)$ has a derivative; for each $\varphi_0 \in W(\alpha, \beta)$, $f_{\alpha,\beta}$ is a linear function of the "increment" u of φ, with $u \in W(0, 0)$. It is evident that

$$f'_{\alpha,\beta}(\varphi_0) \cdot u = f'(\varphi_0) \cdot u \quad \text{for} \quad u \in W(0, 0).$$

In other words, the derivative $f'_{\alpha,\beta}(\varphi_0)$ is the element of $\mathscr{L}(W(0, 0); \mathbf{R})$ obtained from $f'(\varphi_0) \in \mathscr{L}(V; \mathbf{R})$ by restriction to the vector subspace $W(0, 0)$ of V.

By Prop. 8.1.1 of Chap. 1 of *Differential Calculus*, a necessary condition for $f_{\alpha,\beta}$ to possess a relative minimum at $\varphi_0 \in W(\alpha, \beta)$ is that

$$f'(\varphi_0) \cdot u = 0 \quad \text{for all} \quad u \in W(0, 0);$$

in other words, $f'(\varphi_0)$ *should vanish on all vectors* $u \in V$ *such that* $u(a) = 0$ *and* $u(b) = 0$.

DEFINITION. When this is the case we say that φ_0 realizes an *extremum* of $f_{\alpha,\beta}$; we also say that the curve $\varphi: I \to E$ is an *extremal* for the integral

$$\int_a^b F(t, \varphi(t), \varphi'(t))\, dt$$

amongst the curves $\varphi: I \to E$ such that $\varphi(a) = \alpha$ and $\varphi(b) = \beta$.

We shall not consider the question of determining under what conditions an extremal produces a relative *minimum*.

Relation (1.2.3) enables us to give

THEOREM 1.4.1. *In order that* $\varphi \in W(\alpha, \beta)$ *be an extremal, it is necessary and sufficient that*

$$\int_a^b \left[\frac{\partial F}{\partial x}(t, \varphi(t), \varphi'(t)) \cdot u(t) + \frac{\partial F}{\partial y}(t, \varphi(t), \varphi'(t)) \cdot u'(t)\right] dt = 0 \tag{1.4.1}$$

for every function $u\colon I \to E$ *of class* C^1 *such that* $u(a) = 0$ and $u(b) = 0$.

1.5 *Transformation of the condition for an extremum*

We shall effect a transformation of the condition of Theorem 1.4.1. Given the function φ, $(\partial F/\partial x)(t, \varphi(t), \varphi'(t)) = A(t)$ is a known *continuous* function, with values in $\mathscr{L}(E; \mathbf{R})$; similarly, $(\partial F/\partial y)(t, \varphi(t), \varphi'(t)) = B(t)$ is a known *continuous* function with values in $\mathscr{L}(E; \mathbf{R})$. The problem is the following: what condition must be satisfied by the functions $A(t)$ and $B(t)$ in order that

$$\int_a^b (A(t) \cdot u(t) + B(t) \cdot u'(t))\, dt = 0 \tag{1.5.1}$$

for arbitrary $u\colon I \to E$ *of class* C^1, *vanishing for* $t = a$ *and for* $t = b$?

Here is the answer:

THEOREM 1.5.1. *In order that this be the case, it is necessary and sufficient that* $B(t)$ *possess a derivative* $B'(t)$ *equal to* $A(t)$.

This will be proved shortly. First let us apply it to a case of interest; that in which

$$A(t) = \frac{\partial F}{\partial x}(t, \varphi(t), \varphi'(t)), \qquad B(t) = \frac{\partial F}{\partial y}(t, \varphi(t), \varphi'(t)).$$

The result is:

THEOREM 1.5.2. *In order that* $\varphi \in W(\alpha, \beta)$ *be an extremal, it is necessary and sufficient that* $(\partial F/\partial y)(t, \varphi(t), \varphi'(t))$ *be a differentiable function of* t, *and that*

$$\boxed{\frac{d}{dt}\frac{\partial F}{\partial y}(t, \varphi(t), \varphi'(t)) = \frac{\partial F}{\partial x}(t, \varphi(t), \varphi'(t))} \tag{1.5.2}$$

for every $t \in [a, b]$.

The equation (1.5.2) is called *Euler's equation*; it is the equation determining the extremals.

We shall now prove Theorem 1.5.1. It is evident that the condition $A(t) = B'(t)$ is *sufficient*, for we have

$$\int_a^b (B'(t) \cdot u(t) + B(t) \cdot u'(t))\, dt = \int_a^b \frac{d}{dt}(B(t) \cdot u(t))\, dt = B(b) \cdot u(b) - B(a) \cdot u(a) = 0,$$

since $u(a) = u(b) = 0$. It remains to show that the condition is *necessary*.

We shall give two proofs of this necessity: the first, and easiest, is not quite sufficient, since it presupposes the existence of dB/dt. In the second, the existence of dB/dt is effectively established.

FIRST PROOF. Since we suppose that dB/dt exists, the function $t \to B(t) \cdot u(t)$ is differentiable. The left-hand side of (1.5.1) can thus be transformed by integration by parts: we obtain

$$[B(t) \cdot u(t)]_a^b + \int_a^b (A(t) - B'(t)) \cdot u(t)\, dt = 0,$$

where $[B(t) \cdot u(t)]_a^b$ denotes $B(b) \cdot u(b) - B(a) \cdot u(a)$. Now this is zero, since we are assuming that the function $u: I \to E$ vanishes for $t = a$ and $t = b$. The condition sought is therefore:

$$\int_a^b (A(t) - B'(t)) \cdot u(t)\, dt = 0 \tag{1.5.3}$$

for every function $u: I \to E$ *of class* C^1, *such that* $u(a) = 0$ *and* $u(b) = 0$. Thus it suffices to establish the lemma:

Lemma 1.5.3. Let $C: I \to \mathscr{L}(E; \mathbf{R})$ be a continuous function such that

$$\int_a^b C(t) \cdot u(t)\, dt = 0$$

for every function $u: I \to E$ of class C^1, vanishing at $t = a$ and $t = b$. Then C is *identically zero.*

PROOF OF LEMMA 1.5.3. Let us suppose that C does not vanish identically and obtain a contradiction. There exists t_0 such that

$$a < t_0 < b \quad \text{and} \quad C(t_0) \neq 0.$$

Since $C(t_0) \in \mathscr{L}(E; \mathbf{R})$ and $\neq 0$, there exists $u_0 \in E$ such that $C(t_0) \cdot u_0 \neq 0$. Suppose for example

$$C(t_0) \cdot u_0 > 0$$

(if it is < 0, replace u_0 by $-u_0$). Since $C(t)$ is a continuous function of t it follows that

$$C(t) \cdot u_0 > 0$$

for $|t - t_0| \leqslant \epsilon$ (for some $\epsilon > 0$, such that $a \leqslant t_0 - \epsilon \leqslant t_0 + \epsilon \leqslant b$). Now there exists a function $\lambda: I \to \mathbf{R}^+$ whose support is contained in $[t_0 - \epsilon, t_0 + \epsilon]$, and which is > 0 in $]t_0 - \epsilon, t_0 + \epsilon[$ (cf. Chap. 1, § 4.1, Lemma 1). For the function $u: I \to E$ take that function defined by $u(t) = \lambda(t) u_0$ (product of a vector $u_0 \in E$ with the scalar $\lambda(t)$). Then

$$C(t) \cdot u(t) \geqslant 0 \quad \text{for all} \quad t \in I,$$

and

$$C(t) \cdot u(t) > 0 \quad \text{for} \quad t_0 - \epsilon < t < t_0 + \epsilon.$$

Thus the integral $\int_a^b C(t) \cdot u(t)\, dt$ is > 0, contrary to hypothesis; this is the contradiction we set out to establish, and proves Lemma 1.5.3.

SECOND PROOF (complete). Let $A_1(t)$ be the primitive of $A(t)$ which vanishes at $t = 0$:

$$A_1(t) = \int_0^t A(\tau)\, d\tau.$$

Then

$$A(t)\cdot u(t) = A_1'(t)\cdot u(t) = \frac{d}{dt}(A_1(t)\cdot u(t)) - A_1(t)\cdot u'(t).$$

Hence

$$\int_a^b (A(t)\cdot u(t) + B(t)\cdot u'(t))\, dt = [A_1(t)\cdot u(t)]_a^b + \int_a^b (B(t) - A_1(t))\cdot u'(t)\, dt.$$

We require that this be zero for every function $u\colon I \to E$ which vanishes at $t = a$ and $t = b$. The function u' is therefore a *continuous* function $v\colon I \to E$ such that

$$\int_a^b v(t)\, dt = 0; \tag{1.5.4}$$

and conversely, such a v is the derivative of a function u such that $u(a) = 0$ and $u(b) = 0$. The condition sought is therefore

$$\int_a^b (B(t) - A_1(t))\cdot v(t)\, dt = 0 \tag{1.5.5}$$

for every continuous function $v\colon I \to E$ which satisfies (1.5.4). We shall prove the following lemma:

Lemma 1.5.4. Let $D\colon I \to \mathscr{L}(E; \mathbf{R})$ be a continuous function such that

$$\int_a^b D(t)\cdot v(t)\, dt = 0$$

for every continuous function $v\colon I \to E$ such that $\int_a^b v(t)\, dt = 0$. Then D is a *constant.*

If this is proved, then in our case it shows that

$$A_1(t) = B(t) + \text{const.},$$

which means that $B(t)$ possesses a derivative equal to $A(t) = A_1'(t)$. This completes the second proof of Theorem 1.5.1.

It remains to prove Lemma 1.5.4. Again let us obtain a contradiction by supposing that $D(t)$ is not constant: there exist t_1 and t_2 such that

$$a < t_1 < t_2 < b, \qquad D(t_1) \neq D(t_2).$$

Let $u_0 \in E$ such that $D(t_1)\cdot u_0 \neq D(t_2)\cdot u_0$; we can suppose that $D(t_1)\cdot u_0 > D(t_2)\cdot u_0$. Let α_1 and α_2 be in $\mathbf{R}$ such that $D(t_1)\cdot u_0 > \alpha_1 > \alpha_2 > D(t_2)\cdot u_0$. For $\epsilon > 0$ sufficiently small, we have

$$\begin{cases} D(t)\cdot u_0 > \alpha_1 & \text{for} \quad |t - t_1| \leqslant \epsilon \\ D(t)\cdot u_0 < \alpha_2 & \text{for} \quad |t - t_2| \leqslant \epsilon, \end{cases} \tag{1.5.6}$$

and we may suppose that ϵ is so small that

$$a \leqslant t_1 - \epsilon \leqslant t_1 + \epsilon \leqslant t_2 - \epsilon \leqslant t_2 + \epsilon \leqslant b.$$

Let $\lambda: \mathbf{R} \to \mathbf{R}^+$ be a function of class C^∞, vanishing for $t \leqslant -\epsilon$ and for $t \geqslant \epsilon$, and such that $\lambda(t) > 0$ for $-\epsilon < t < \epsilon$ (cf. Chap. 1, § 4.1). The function

$$\mu(t) = \lambda(t - t_1) - \lambda(t - t_2)$$

is of class C^∞, satisfies $\int_a^b \mu(t)\, dt = 0$, is > 0 for $t_1 - \epsilon < t < t_1 + \epsilon$, is < 0 for $t_2 - \epsilon < t < t_2 + \epsilon$, and vanishes elsewhere.

Put

$$v(t) = \mu(t) \cdot u_0;$$

this is a continuous function $I \to E$; we have $\int_a^b v(t)\, dt = 0$, and further that

$$\int_a^b D(t) \cdot v(t)\, dt = \int_{t_1 - \varepsilon}^{t_1 + \varepsilon} \lambda(t - t_1)(D(t) \cdot u_0)\, dt - \int_{t_2 - \varepsilon}^{t_2 + \varepsilon} \lambda(t - t_2)(D(t) \cdot u_0)\, dt.$$

By the inequalities (1.5.6), we have

$$\begin{aligned} \int_a^b D(t) \cdot v(t)\, dt &> \alpha_1 \int_{t_1 - \varepsilon}^{t_1 + \varepsilon} \lambda(t - t_1)\, dt - \alpha_2 \int_{t_2 - \varepsilon}^{t_2 + \varepsilon} \lambda(t - t_2)\, dt \\ &> (\alpha_1 - \alpha_2) \int_{-\varepsilon}^{+\varepsilon} \lambda(t)\, dt > 0, \end{aligned}$$

contrary to hypothesis. This contradiction establishes the lemma.

This completes the proof of Theorem 1.5.1.

1.6 *Calculation of $f'(\varphi) \cdot u$ for an extremal*

In the preceding section we obtained the *Euler equation* for the extremals φ from the condition

$$f'(\varphi) \cdot u = 0$$

for every function $u: I \to E$, of class C^1, *vanishing for $t = a$ and for $t = b$.*

Let us now abandon the hypotheses $u(a) = 0$ and $u(b) = 0$. With the notation of § 1.5 we have

$$f'(\varphi) \cdot u = \int_a^b (A(t) \cdot u(t) + B(t) \cdot u'(t))\, dt;$$

we have seen that, for an extremal, it is true that $A(t) = B'(t)$, therefore,

$$f'(\varphi) \cdot u = B(b) \cdot u(b) - B(a) \cdot u(a)$$

(which is indeed zero if $u(a) = u(b) = 0$). Thus,

$$(1.6.1) \qquad \boxed{f'(\varphi) \cdot u = \frac{\partial F}{\partial y}(b, \varphi(b), \varphi'(b)) \cdot u(b) - \frac{\partial F}{\partial y}(a, \varphi(a), \varphi'(a)) \cdot u(a)}$$

each time that $\varphi: I \to E$ *is an extremal.* This formula shows how $f'(\varphi)\cdot u$ depends on the function u; in fact, it depends only on the values of this function at $t = a$ and $t = b$.

If φ is not an extremal, then, in general, the expression for $f'(\varphi)\cdot u$ will involve the function $u(t)$ for all values of $t \in [a, b]$. In fact, we have (as in § 1.5)

$$(1.6.2) \qquad f'(\varphi)\cdot u = \frac{\partial F}{\partial y}(b, \varphi(b), \varphi'(b))\cdot u(b) - \frac{\partial F}{\partial y}(a, \varphi(a), \varphi'(a))\cdot u(a)$$

$$+ \int_a^b \left[\frac{\partial F}{\partial x}(t, \varphi(t), \varphi'(t)) - \frac{d}{dt}\frac{\partial F}{\partial y}(t, \varphi(t), \varphi'(t))\right]\cdot u(t)\, dt,$$

at least provided that $\varphi: I \to E$ is such that the function $(\partial F/\partial y)(t, \varphi(t), \varphi'(t))$ of the variable t possesses a continuous derivative.

2. Study of the Euler equation. Existence of extremals. Examples

2.1 *The Euler equation for the case* $E = \mathbf{R}^n$

A point x (resp. y) of E is now defined by n coordinates $x_1, \ldots, x_n$ (resp. $y_1, \ldots, y_n$) which are real numbers. The function

$$F(t, x_1, \ldots, x_m, y_1, \ldots, y_n)$$

is defined and of class C^k ($k \geqslant 1$) in an open $U \subset \mathbf{R}^{2n+1}$. A curve $\varphi: I \to \mathbf{R}^n$ of class C^1 is defined by n functions $\varphi_i(t)$ of class C^1; and φ belongs to an open Ω if

$$(t, \varphi_1(t), \ldots, \varphi_n(t), \varphi_1'(t), \ldots, \varphi_n'(t)) \in U$$

for all $t \in I$. We associate with φ the integral

$$f(\varphi) = \int_a^b F(t, \varphi_1(t), \ldots, \varphi_n(t), \varphi_1'(t), \ldots, \varphi_n'(t))\, dt.$$

The Euler equation (1.5.2) which characterizes the extremals is here equivalent to the n scalar equations

$$(2.1.1) \qquad \frac{d}{dt}\left(\frac{\partial F}{\partial y_i}\right) = \frac{\partial F}{\partial x_i}, \qquad 1 \leqslant i \leqslant n,$$

where $\partial F/\partial x_i$ denotes the function

$$\frac{\partial F}{\partial x_i}(t, \varphi_1(t), \ldots, \varphi_n(t), \varphi_1'(t), \ldots, \varphi_n'(t)),$$

and similarly for $\partial F/\partial y_i$.

We shall write down the explicit form of this system of equations, *under the assumption that* F *is of class* C^2, *and that the extremal considered is of class* C^2 (it will be seen shortly (cf. Prop. 2.1.1) that, when F satisfies certain conditions, *all* the extremals are effectively of class C^2). Since $\partial F/\partial y_i$ is a function of t (the x_i being replaced by $\varphi_i(t)$, and the y_i by $\varphi_i'(t)$), we obtain

$$\frac{d}{dt}\left(\frac{\partial F}{\partial y_i}\right) = \frac{\partial^2 F}{\partial t \partial y_i}(t, \varphi(t), \varphi'(t)) + \sum_{j=1}^{n} \frac{\partial^2 F}{\partial x_j \partial y_i}(t, \varphi(t), \varphi'(t)) \cdot \varphi_j'(t)$$
$$+ \sum_{j=1}^{n} \frac{\partial^2 F}{\partial y_j \partial y_i}(t, \varphi(t), \varphi'(t)) \cdot \varphi_j''(t).$$

Thus the Euler equations have now been expressed as a system of n second-order differential equations for the unknown functions $x_1, \ldots, x_n$ of t (we shall write x_i' in place of y_i):

$$\frac{\partial^2 F}{\partial t \partial x_i'}(t, x, x') + \sum_{j=1}^{n} \frac{\partial^2 F}{\partial x_j \partial x_i'}(t, x, x') \cdot x_j' + \sum_{j=1}^{n} \frac{\partial^2 F}{\partial x_j' \partial x_i'}(t, x, x') \cdot x_j''$$
$$= \frac{\partial F}{\partial x_i}(t, x, x'), \quad \text{for} \quad 1 \leqslant i \leqslant n. \tag{2.1.2}$$

Suppose now that the determinant

$$\det\left(\frac{\partial^2 F}{\partial x_i' \partial x_j'}(t, x, x')\right) \tag{2.1.3}$$

is $\neq 0$ for $(t, x, x') \in U$. Then the system (2.1.2) may be expressed in the form

$$x_i'' = G_i(t, x_1, \ldots, x_n, x_1', \ldots, x_n') \qquad (1 \leqslant i \leqslant n),$$

where the G_i are of class C^{k-2} if F is of class C^k.

PROPOSITION 2.1.1. *If the determinant* (2.1.3) *is* $\neq 0$, *and* F *is of class* C^2, *then the extremals*

$$x_i = \varphi_i(t) \qquad (1 \leqslant i \leqslant n)$$

are of class C^2.

PROOF. Write

$$z_i = \frac{\partial F}{\partial y_i}(t, x_1, \ldots, x_n, y_1, \ldots, y_n), \qquad 1 \leqslant i \leqslant n. \tag{2.1.4}$$

The Jacobian of the n functions $z_1, \ldots, z_n$ with respect to $y_1, \ldots, y_n$ is $\det(\partial^2 F/\partial y_i \partial y_j)$, which by hypothesis $\neq 0$. Therefore, in the neighbourhood of a point $(t, x_1, \ldots, x_n, y_1, \ldots, y_n)$ we may apply the theorem of implicit functions to obtain

$$y_i = G_i(t, x_1, \ldots, x_n, z_1, \ldots, z_n), \qquad 1 \leqslant i \leqslant n, \tag{2.1.5}$$

the G_i being of class C^1. In order to be able to write the Euler equations, we must first of all express y_i as the derivative of x_i with respect to t, then

$$\frac{d}{dt}\left(\frac{\partial F}{\partial y_i}\right) = \frac{\partial F}{\partial x_i}.$$

Therefore the extremals are the solutions of

$$\text{(2.1.6)} \qquad \begin{cases} \dfrac{dx_i}{dt} = G_i(t, x_1, \ldots, x_n, z_1, \ldots, z_n) \\ \dfrac{dz_i}{dt} = \dfrac{\partial F}{\partial x_i}(t, x_1, \ldots, x_n, G_1(t, x, z), \ldots, G_n(t, x, z)) \end{cases}$$

i.e., of a differential system of class C^1 in $2n$ unknown functions x_i and z_i of the variable t. For each extremal the x_i and the z_i are therefore functions of class C^1 of t; by (2.1.5) the y_i are functions of class C^1 of t, and since $y_i = dx_i/dt$, it is clear that the $x_i(t)$ are functions of class C^2 for every extremal.

Q.E.D.

By supposing that F is of class C^2, and that the determinant (2.1.3) is $\neq 0$, we may apply the theorem of local existence and uniqueness of differential systems to the system (2.1.6); *through a point* $(t, x_1, \ldots, x_n)$ *there passes one and only one extremal such that the* dx_i/dt *take given values at* $t = t_0$.

2.2 *Examples*

Preliminary remark: in the space considered in § 2.1 (i.e., $E = \mathbf{R}^n$), suppose that for *one* $i (1 \leqslant i \leqslant n)$ the function $F(t, x_1, \ldots, x_n, y_1, \ldots, y_n)$ is independent of x_i, i.e. that $\partial F/\partial x_i = 0$. The Euler equation (2.1.1) then shows that *on every extremal* $(\partial F/\partial y_i)(t, \varphi(t), \varphi'(t))$ *is constant.* In other words, $(\partial F/\partial y_i)(t, x, y)$ is a *first integral* of the differential system of the extremals.

First example. Let $F = \sqrt{(y_1)^2 + \cdots + (y_n)^2}$. This is the problem of finding the curves in $\mathbf{R}^n$ which are *extremals for the length,* since

$$\int_a^b \sqrt{\varphi_1'(t)^2 + \cdots + \varphi_n'(t)^2}\, dt$$

is none other than the length of arc of the curve $\varphi: I \to \mathbf{R}^n$ of class C^1. In this example $\partial F/\partial x_i = 0$ for all i; therefore, for every i

$$\frac{\partial F}{\partial y_i} = \frac{y_i}{\sqrt{(y_1)^2 + \cdots + (y_n)^2}}$$

is constant on each extremal curve. In other words, the *direction cosines* of the tangent vector of the extremal are constant. This means that the tangent of the extremal has a fixed direction: *the extremals are straight lines in* $\mathbf{R}^n$, as expected. We can use formula (1.6.1) to calculate the "infinitesimal variations" of the length of a straight segment as a function of the "infinitesimal variations" of its end-points: in this formula, the vectors $u(a)$ and $u(b)$ are the infinitesimal displacements of the extremities A and B of the straight segment (corresponding to the values a and b of the parameter t); the right-hand side of (1.6.1) is equal to the difference between the orthogonal projections, on the segment $\overline{AB}$, of the vectors $u(a)$ and $u(b)$ ("formula of Joseph Bertrand").

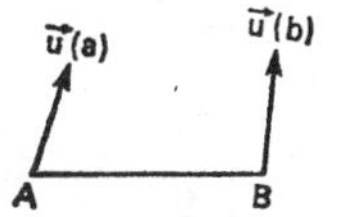

Second example. Suppose, more generally, that

$$F = \alpha(x)\sqrt{(y_1)^2 + \cdots + (y_n)^2},$$

where $\alpha(x)$ is a function of $x_1, \ldots, x_n$ of class C^k (k as large as required). If α is independent of *one* of the variables x_i, then

$$\frac{\partial F}{\partial y_i} = \alpha(x) \frac{y_i}{\sqrt{(y_1)^2 + \cdots + (y_n)^2}}$$

is *constant* along each extremal: the projection onto the ith coordinate axis of the vector of length $\alpha(x)$ in the direction of the tangent to the extremal is constant. This problem in the calculus of variations occurs notably in theory of light propagating in an *isotropic* medium; "isotropic" signifies that, at each point, the velocity of light is the same in all directions; it is inversely proportional to the *refractive index* of the medium at that point. The refractive index is a function of position (>0), that may be supposed sufficiently differentiable, and *independent of the time*. The integral $\int F\,dt$ may be interpreted as $\int ds/v$, ds being the element of arc of the ray under consideration, and v being the velocity of light at the point considered on the ray; it is therefore equal to the *time of passage of the light along the ray*. Now Fermat's principle says that the trajectories of the light minimize the time of passage. Therefore the trajectories (rays) are the extremals of the problem. The law found above says that if the refractive index is independent of x_i, say, the projection on the x_i-axis, of the vector tangential to the light ray and whose length is equal to the refractive index at the point considered, is *constant* along the trajectory of the light. From this, it is easy to derive (by passage to the limit) Descartes' *law of refraction* when the medium is divided into two homogeneous regions by a plane P: if n_1 and n_2 denote the refractive indices, and if i_1 and i_2 denote the angles between the light ray and the plane P, then

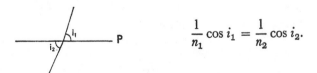

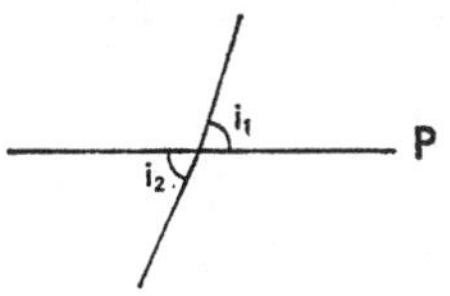

$$\frac{1}{n_1}\cos i_1 = \frac{1}{n_2}\cos i_2.$$

Here is a particularly interesting case of our second example, for $n = 2$, with

(2.2.1) $$\alpha(x) = (x_2)^k,$$

and we are concerned with the *half-plane* $x_2 > 0$ (with coordinates x_1, x_2), k being an exponent. Then

$$(x_2)^k \frac{x_1'}{\sqrt{x_1'^2 + x_2'^2}} = \text{const.}$$

on each extremal curve. This effectively determines the extremals. We give a few other cases which the reader may consider as exercises.

Case where $k = -1$; this gives the integral $\int ds/x_2$ in the half-plane $x_2 > 0$. This is the case of Lobatchewski geometry in the "half-plane of Poincaré". The extremals are arcs of circles centred on axis of x_1; they make the integral a *minimum.*

Case where $k = +1$; here we have the integral $\int x_2\, ds$. The extremals are the "catenaries" $x_2 = a \cosh (x_1/a + \text{const.})$. This problem arises in the search for surfaces of revolution of minimum area.

Case where $k = -\frac{1}{2}$; this gives the integral $\int ds/\sqrt{x_2}$. This is the problem of determining those curves C in the half-plane $x_2 > 0$ such that a point mass under the action of gravity (the x_2-axis being in the direction of gravity) which slides on C without friction and with a speed $v = \sqrt{2gx_2}$, takes the least time to descend. The extremals are the arcs of a cycloid.

Case where $k = \frac{1}{2}$; $\int \sqrt{x_2}\, ds$. This problem occurs in the theory of the parabolic motion of a projectile under the action of gravity acting in the direction of the x_2-axis, the initial speed being equal to $\sqrt{2gx_2}$.

2.3 *The Lagrange equations in mechanics*

We shall consider a material system whose diverse possible configurations are defined by a finite number of (scalar) parameters $q_1, \ldots, q_n$. For example, a finite number of free point masses; more generally, the system will consist of a finite number of rigid bodies subject to "holonomic" constraints (the configuration of a finite system of rigid bodies is described by a finite number of parameters: if there exist "constraints" between the bodies, then the parameters must satisfy certain "relations"; these may be relations between differentials, equalities or inequalities of functions of the parameters. The "holonomic" case is that in which, locally, the constraints may be expressed by saying that certain of the parameters are known functions of the others, supposed independent).

If the parameters q_i are differentiable functions of the time t, the *kinetic energy* T of the system at the instant t may be expressed as a function of $t, q_1, \ldots, q_n$ and the derivatives $q_i' = dq_i/dt$; for fixed $t, q_1, \ldots, q_n$, this function is *quadratic* in $q_1', \ldots, q_n'$. In good cases it is a *homogeneous* function of the second degree in the q_i'; in general, we have

$$T = T_2 + T_1 + T_0,$$

where T_2 is a homogeneous polynomial of degree 2, T_1 is a linear (homogeneous) function, and T_0 is independent of the q_i'.

We suppose that at each instant the material system is subject to a system of forces defined by a *potential function* $U(t, q_1, \ldots, q_n)$: this means that for t fixed, the "work" done by the system of forces when the configuration of the system is changed from $(q_1^0, \ldots, q_n^0)$ to $(q_1^1, \ldots, q_n^1)$ is equal to $U(q_1^1, \ldots, q_n^1) - U(q_1^0, \ldots, q_n^0)$. Then the fundamental principle of mechanics states that *the time evolution of the material system is defined by a curve* $t \mapsto (q_1(t), \ldots, q_n(t))$ *which is an extremal for the integral*

$$\int (T + U)\, dt.$$

This is called the "action integral", and the principle is known as Hamilton's "*principle of least action*".

Thus the equations describing the evolution of the material system are the Euler equations for this problem. Here, $F = T + U$, the variables $x_1, \ldots, x_n$ are the configuration parameters $q_1, \ldots, q_n$. Taking account of the fact that U is independent of $q'_1, \ldots, q'_n$, the Euler equations become

$$\boxed{\frac{d}{dt}\frac{\partial T}{\partial q'_i} = \frac{\partial T}{\partial q'_i} + \frac{\partial U}{\partial q_i}} \qquad 1 \leqslant i \leqslant n \tag{2.3.1}$$

These are the classical "Lagrange equations" of mechanics.

2.4 *More about the general case: the case where* $F(t, x, y)$ *is independent of* t

We determine the extremals of $\int F(x, y)\, dt$, where F is of class C^1 in an open $U \subset E \times E$.

THEOREM 2.4.1. If F is independent of t, the function $(\partial F/\partial y) . y - F$ (which has values in $\mathbf{R}$) is *constant* along each extremal.

PROOF. Of course, $(\partial F/\partial y)(x, y) \cdot y$ denotes the value of the element $(\partial F/\partial y)(x, y) \in \mathscr{L}(E; \mathbf{R})$ on the vector $y \in E$. We must therefore verify that the derivative with respect to t of the composite function

$$\frac{\partial F}{\partial y}(\varphi(t), \varphi'(t))\varphi'(t) - F(\varphi(t), \varphi'(t))$$

vanishes when $t \to \varphi(t)$ is an extremal curve. By the formula for the derivative of a bilinear function (which is applied to $(\partial F/\partial y) \cdot y$ as a bilinear function of $\partial F/\partial y$ and y), we have

$$\frac{d}{dt}\left(\frac{\partial F}{\partial y} \cdot y\right) - \frac{dF}{dt} = \frac{d}{dt}\left(\frac{\partial F}{\partial y}\right) \cdot y + \frac{\partial F}{\partial y} \cdot \frac{dy}{dt} - \frac{\partial F}{\partial x} \cdot \frac{dx}{dt} - \frac{\partial F}{\partial y} \cdot \frac{dy}{dt}, \tag{2.4.1}$$

where x is replaced by $\varphi(t)$, y by $dx/dt = \varphi'(t)$; on the other hand,

$$\frac{d}{dt}\left(\frac{\partial F}{\partial y}\right) = \frac{\partial F}{\partial x} \qquad \text{(Euler equation)}$$

so that the right-hand side of (2.4.1) is zero.

Q.E.D.

When F is a homogeneous polynomial of degree n in y (for each x), we have

$$\frac{\partial F}{\partial y}\cdot y = nF; \tag{2.4.2}$$

this is the classical identity of Euler, which is obtained by writing

$$F(x, \lambda y) = \lambda^n F(x, y)$$

and using the condition that the derivatives of both sides with respect to λ are equal for $\lambda = 1$.

Example. If, in mechanics, U and T are *independent of t*, and if we put

$$T = T_2 + T_1 + T_0,$$

where T_2 is homogeneous of degree 2 in the q_i', T_1 homogeneous of degree 1 in the q_i', and T is independent of the q_i', then

$$\sum_{i=1}^{n} \frac{\partial F}{\partial q_i'} q_i' - F = (2T_2 + T_1) - (T_2 + T_1 + T_0 + U) = T_2 - T_0 - U.$$

This leads to the *energy theorem* (in the generalized form due to Painlevé): if T and U do not depend on t, $T_2 - T_0 - U$ is constant on each trajectory.

2.5 *The case where* $F(x, y)$ *is a homogeneous quadratic function of* y

Theorem 2.4.1 tells us that $F(\varphi(t), \varphi'(t))$ is *constant along each extremal.* Let us now suppose that, for each x, $F(x,y)$ is a *positive non-degenerate* quadratic form in y (cf. *Differential Calculus,* Chap. 1, § 8). Then we can consider the function $G(x, y) = \sqrt{F(x, y)}$; if F is of class C^k, the same is true for $G(x, y)$ at every point (x, y) such that $y \neq 0$. Therefore, apart from the variational problem defined by the integral $\int F(x, y)\, dt$, we may also consider that defined by $\int \sqrt{F(x, y)}\, dt$. We propose to compare the extremals of the two problems.

However, it is first necessary to distinguish between a "parametrized curve" and a "geometrical curve". A *parametrized curve* of the space E is simply a function

$$\varphi: I \to E$$

of a real variable $t \in I = [a, b]$, which we suppose to be of class C^1; we suppose further that, for $t \in I$, $\varphi'(t) \neq 0$, so that the point $(\varphi(t), \varphi'(t))$ is indeed in the open U which, by hypothesis, contains no point (x, y) such that $y = 0$. On such a parametrized curve we can make a change of parameter

$$t = \lambda(u),$$

where λ is a strictly increasing mapping of class C^1 of $[a, b]$ onto $[a', b']$, with derivative $\lambda'(u) \neq 0$ for all u; we obtain a new parametrized curve $u \to \varphi(\lambda(u))$. By definition, a *geometrical curve* is a class of parametrized curves which is obtained from one of them by change of parameter as above.

Having said this, it is clear that if two parametrized curves belong to the same class, the integral

$$\int \sqrt{F(\varphi(t), \varphi'(t))}\, dt$$

has the same value for each of the curves; this follows from the fact that

$$\sqrt{F(x, \xi y)} = \xi\sqrt{F(x, y)} \quad \text{for all} \quad \xi > 0.$$

Thus the integral

$$\int_a^b \sqrt{F(\varphi(t), \varphi'(t))}\, dt$$

is in fact completely determined by a geometrical curve. The extremals are therefore geometrical curves: the parametrization of the curve is of no importance. Of course, this is not true for $\int_a^b F(\varphi(t), \varphi'(t))\, dt$. The relation between the two problems is clarified by the two following propositions:

PROPOSITION 2.5.1. *Every extremal of* $\int F\, dt$ *is also an extremal of* $\int \sqrt{F}\, dt$.

In fact, the extremals of the first problem are the solutions of

$$\frac{d}{dt}\left(\frac{\partial F}{\partial y}\right) = \frac{\partial F}{\partial x} \tag{2.5.1}$$

whereas the extremals of the second problem are the solutions of

$$\frac{d}{dt}\left(\frac{1}{\sqrt{F}}\frac{\partial F}{\partial y}\right) = \frac{1}{\sqrt{F}}\frac{\partial F}{\partial x}. \tag{2.5.2}$$

But along each extremal (2.5.1) we know that F is constant; therefore this extremal satisfies (2.5.2).

Q.E.D.

PROPOSITION 2.5.2. *Every geometrical extremal of* $\int \sqrt{F}\, dt$ *possesses a unique parametrization* (by $t \in [a, b]$) *such that* $F(\varphi(t), \varphi'(t))$ *is constant. This parametrized curve is therefore also an extremal of* $\int F\, dt$.

PROOF. Consider an extremal

$$x = \psi(u)$$

of $\int \sqrt{F}\, dt$. We search for $u = \lambda(t)$ such that

$$\sqrt{F(\psi(\lambda(t)), \psi'(\lambda(t)))} \cdot \lambda'(t)$$

is independent of t; now this is equal to

$$\lambda'(t) \cdot \sqrt{F(\psi(u), \psi'(u))}, \quad \text{with} \quad u = \lambda(t).$$

The function $\sqrt{F(\psi(u), \psi'(u))}$ is a known continuous function $f(u) > 0$; we require that

$$\frac{du}{dt}\cdot f(u) = c \quad (c \text{ constant}),$$

which gives $t = (1/c) \int f(u)\, du + k$, where k is constant.

We determine c and k in such a manner that $t = a$ for $u = a$, and $t = b$ for $u = b$. Thus t is a known function of u, with derivative > 0; hence u is a known function of t, which gives the required change of parameter. With this parameter t, the curve $x = \varphi(t) = \psi(\lambda(t))$ is again a solution of (2.5.2).

Q.E.D.

2.6 *Geodesics on a variety*

The length of an element of a curve lying in a surface in $\mathbf{R}^3$ has been defined in Chap. 1, § 4.12. More generally, let M be a variety of dimension p and of class C^k in $\mathbf{R}^n$; if we use a local parametrization of M, the coordinates $x_1, \ldots, x_n$ of a point of M are functions of class C^2 of p parameters $u_1, \ldots, u_p$ which vary in an open $U \subset \mathbf{R}^p$, the matrix $\{\partial x_i/\partial u_j\}$ being of rank p at every point of U. When $u_1, \ldots, u_p$ are functions of class C^1 of a parameter $t \in [a, b]$, whose derivatives $u_i'(t)$ are not all simultaneously null, the composite functions

$$x_i(u_1(t), \ldots, u_p(t)) \qquad (1 \leqslant i \leqslant n)$$

define a curve of class C^1 in $\mathbf{R}^n$, whose length element is

$$ds = \sqrt{\left(\frac{dx_1}{dt}\right)^2 + \cdots + \left(\frac{dx_n}{dt}\right)^2}\, dt.$$

If dx_i/dt is calculated by means of the formula which gives the derivative of a composite function, we obtain

$$\begin{aligned}\left(\frac{ds}{dt}\right)^2 &= \sum_{i=1}^{n} \left(\sum_{j=1}^{p} \frac{\partial x_i}{\partial u_j} u_j'\right)^2, \quad \text{with} \quad u_j' = \frac{du_j}{dt} \\ &= F(u_1, \ldots, u_p, u_1', \ldots, u_p'),\end{aligned}$$

where F is a positive non-degenerate, homogeneous quadratic form in $u_1', \ldots, u_p'$, whose coefficients are functions of class C^{k-1} of $u_1, \ldots, u_p$. The length element of a parametrized curve traced on M is therefore

$$\sqrt{F(u_1, \ldots, u_p, u_1', \ldots, u_p')}\, dt, \tag{2.6.1}$$

where the u_i are replaced by the functions $u_i(t)$ which define the curve, and the u_i' are replaced by their derivatives.

By definition, the *geodesics* of the variety M are the *extremal* curves of

$$\int \sqrt{F(u_1, \ldots, u_p, u_1', \ldots, u_p')}\, dt$$

These are "geometrical curves", since the integral gives the length of arc of the curve, which is unchanged in value when a change of parameter is effected.

The results of § 2.5 may be applied to geodesics. We are thus led to consider the extremals of

$$\int F(u_1, \ldots, u_p, u_1', \ldots, u_p')\, dt, \tag{2.6.2}$$

where the value of the integral depends on the choice of the parametrization of the geometrical curve. The extremals of (2.6.2) are therefore the geodesics defined by a parameter t, such that $F(u_1, \ldots, u_p, u_1', \ldots, u_p')$ is *constant* on the curve; in other words, *the parameter t must be proportional to the length of arc of the curve considered.* Thus: *the extremals of* (2.6.2) *are geodesics with parameter proportional to the arc length.*

When searching for geodesics, it is better to determine the extremals of (2.6.2): this avoids radicals, and at the same time, gives the length to within a constant factor.

Let us obtain the explicit form of the Euler equations (2.6.2); to do this, we use vector notation: u is the function $I \to \mathbf{R}^p$ of class C^1, u' is its derivative, and we write $F(u, u')$ for $F(u_1, \ldots, u_p, u_1', \ldots, u_p')$. Then

$$\frac{d}{dt}\left(\frac{\partial F}{\partial u'}\right) = \frac{\partial F}{\partial u}, \quad \text{i.e.,}$$

$$\frac{\partial^2 F}{\partial u \cdot \partial u'} \cdot u' + \frac{\partial^2 F}{\partial u' \cdot \partial u'} \cdot u'' = \frac{\partial F}{\partial u}, \tag{2.6.3}$$

where u' and u'' denote the first and second derivatives of the unknown function u of t. Both sides of this equation have values in $\mathscr{L}(\mathbf{R}^p; \mathbf{R})$, the dual of the space $E = \mathbf{R}^p$ in which the values of $u(t)$ are situated. Note that $\partial^2 F/(\partial u \partial u')$ takes values in the space of bilinear mappings $E \times E \to \mathbf{R}$, and that the notation $\partial^2 F/(\partial u \partial u')$ is ambiguous, since there are two ways of associating a vector $u' \in E$, and a bilinear mapping $E \times E \to \mathbf{R}$ with a linear mapping $E \to \mathbf{R}$. The ambiguity is removed by recalling that $\partial^2 F/(\partial u \partial u') \cdot u'$ denotes the value, on the vector u', of the derivative $\partial/\partial u$ of the function $\partial F/\partial u'$ (not to be confused with the value, on u', of the derivative $\partial/\partial u'$ of the function $\partial F/\partial u$).

Let us also note that, F being homogeneous of degree 2, in u', $\partial F/\partial u'$ is homogeneous of degree 1 in u', as is its derivative $(\partial/\partial u)(\partial F/\partial u')$; and $\partial^2 F/(\partial u' \partial u')$ is homogeneous of degree 0, i.e. independent of u'; as for $\partial F/\partial u$, it is homogeneous of degree 2 in u'. Finally the differential equation for the parametrized geodesics is written

$$\frac{\partial^2 F}{\partial u' \partial u'}(u) \cdot u'' = G(u, u'), \tag{2.6.4}$$

where G has values in $\mathscr{L}(E; \mathbf{R})$, is homogeneous of degree 2 in u', and is of class C^{k-1} in u.

Since F is a non-degenerate quadratic form in u', the mapping

$$\xi \mapsto \frac{\partial^2 F}{\partial u' \cdot \partial u'}(u) \cdot \xi$$

is, for each u, an *isomorphism* of E onto its dual $\mathscr{L}(E; \mathbf{R})$; let $\sigma(u) : \mathscr{L}(E; \mathbf{R}) \to E$ be the inverse isomorphism, which is of class C^k in u. We obtain an equation equivalent to (2.6.4) by applying $\sigma(u)$ to both sides of (2.6.4):

$$u'' = \sigma(u)\cdot G(u, u'). \tag{2.6.5}$$

The right-hand side $H(u, u')$ *is homogeneous of degree* 2 *in* u', *of class* C^{k-1} *in* u, *with values in* E. This is the differential equation of the geodesics when the parameter is proportional to the length.

Let us now apply the *theorem of local existence and uniqueness for differential equations of the second order* to equation (2.6.5). But first, we can verify that if $u(t)$ is a solution, so is $u(ct)$ for arbitrary constant c: this is already known to be true, since t is proportional to the arc length of the geodesic; and in (2.6.5) the right-hand side is the function $H(u, u')$ which is homogeneous quadratic in u'. This said, given the initial value $u(0)$ and the initial derivative $u'(0)$, there exists a unique geodesic $u(t)$ (in an interval $-\epsilon \leqslant t \leqslant +\epsilon$) corresponding to these initial values. If $u'(0)$ is multiplied by a constant > 0, the geometrical curve is unchanged. Thus, *there exists a unique geometrical geodesic passing through a given point and tangential at that point to a given straight line.*

Next, we shall apply Theorem 3.8.1 of Chap. 2 of *Differential Calculus*, valid for every second order differential equation of class C^1, to equation (2.6.5). To do this, we shall therefore suppose that the quadratic form $F(u, u')$ is a function of class C^2 in u, so that the parametrized geodesics will be the solutions of

$$u'' = H(u, u'),$$

where $H(u, u')$ is of class C^1 in u. We shall apply Theorem 3.8.1 of Chap. 2 of *Differential Calculus*, replacing x by u and x' by u'. The solution corresponding to the given initial conditions is

$$u(t) = a \quad \text{(constant)}.$$

This being so, the "note" following Theorem 3.8.1 under consideration enables us to conclude that:

If two points $b, c \in R^p$ *are given sufficiently close to* a, *there exists an initial speed* $u'(0)$ *in the neighbourhood of* 0, *and one only, such that the geodesic* $u(t)$ *for which* $u(0) = b$ *exists for* $|t| \leqslant 1$ *and such that* $u(1) = c$. *In brief: there exists one and only one geodesic which passes through two points* b *and* c *sufficiently close to* a.

2.7 *Extremum problems for curves constrained to lie on a variety*

For simplicity let us suppose $E = \mathbf{R}^n$, and let S be a variety of class C^p (p sufficiently large) in $\mathbf{R}^n$ (cf. Chap. 1, § 4.7). S being given, we shall now suppose that we deal with only those curves of class C^1

$$\varphi: I \to \mathbf{R}^n$$

such that, for all $t \in I = [a, b]$, it is true that $\varphi(t) \in S$; such a curve will be said to be "traced on S". These curves form a subset V_s of the Banach space V of all curves of class C^1 (cf. § 1.1); V_s has the topology of V, but, of course, V_s is not a *vector* subspace.

As before, suppose we are given a numerical function of class C^k in an open $U \in R \times$

$E \times E$, and let the *open* $\Omega \subset V$ denote the set of those φ such that $(t, \varphi(t), \varphi'(t)) \in U$ for every $t \in I$. Then $V_s \cap \Omega$ is an open subset of the topological space V_s.

The minimum problem. Let $\Omega_s(\alpha, \beta)$ be the set of $\varphi \in V_s \cap \Omega$ such that $\varphi(a) = \alpha$ and $\varphi(b) = \beta$; this is a topological space which we suppose to be *non-empty* (which implies that the points $\alpha \in \mathbf{R}^n$ and $\beta \in \mathbf{R}^n$ belong to the same connected component of S). Let $\varphi_0 \in \Omega_s(\alpha, \beta)$. We now determine whether φ_0 gives a *relative minimum* of

$$f(\varphi) = \int_a^b F(t, \varphi(t), \varphi'(t))\, dt \tag{2.7.1}$$

for $\varphi \in \Omega_s(\alpha, \beta)$; in other words, whether

$$f(\varphi) \geqslant f(\varphi_0)$$

for every $\varphi \in \Omega(\alpha, \beta)$ sufficiently close to φ_0 (in the sense of the topology of $\Omega_s(\alpha, \beta)$). We shall prove the following theorem:

THEOREM 2.7.1. In order that φ_0 furnish a relative minimum of (2.7.1) when φ varies over $\Omega_s(\alpha, \beta)$, it is *necessary* that

$$f'(\varphi_0)\cdot u = 0 \tag{2.7.2}$$

for every $u\colon I \to \mathbf{R}^n$ of class C^1 which possesses the following property:

P(S): $u(a) = 0$, $u(b) = 0$, *and, for every* $t \in I$, *the vector* $u(t) \in \mathbf{R}^n$ *is tangential to the variety* S *at the point* $\varphi_0(t)$. (For the definition of the tangent vector of a variety see Chap. 1, Prop. 4.7.2. We say that a vector is tangential to S at a point $x \in S$ if it belongs to the tangential vector space $T_x(S)$).

Note. In § 1.4 it was shown that a necessary condition for a minimum is $f'(\varphi_0)\cdot u = 0$ for every $u\colon I \to \mathbf{R}^n$ of class C^1 which vanishes at $t = a$ and at $t = b$. Theorem 2.7.1 deals with a more restricted set of functions u; thus condition (2.7.2) imposed on φ_0 is much weaker than the condition of § 1.4. This is fortunate, for, considering for example the problem of minimizing the arc length of a curve, the condition of § 2.1 restricts this curve to a straight line. Now in the present case we require that the curve be traced on the given variety S, and in general two points of S cannot be joined by a straight line traced in S.

Before proving Theorem 2.7.1 let us recall the relation

$$f'(\varphi_0)\cdot u = \int_a^b \left[\frac{\partial F}{\partial x}(t, \varphi_0(t), \varphi_0'(t))\cdot u(t) + \frac{\partial F}{\partial y}(t, \varphi(t), \varphi'(t))\cdot u'(t)\right] dt \tag{2.7.3}$$

(cf. relation (1.2.3)).

PROOF OF THEOREM 2.7.1. A function $\psi(t, \lambda)$ of the two real variables t and λ, defined and of class C^1 when

$$t \in I \quad \text{and} \quad |\lambda| \leqslant \epsilon \qquad (\epsilon > 0 \text{ small}),$$

taking values in S, defines a one-parameter family of curves $\varphi_\lambda\colon I \to S$, viz.,

$$\varphi_\lambda(t) = \psi(t, \lambda).$$

Let us suppose that $\psi(t, 0)$ is the function $\varphi_0(t)$ of the statement of Theorem 2.7.1, and that $\varphi_\lambda(a) = a$, $\varphi_\lambda(b) = b$ for all λ; then $\varphi_\lambda \in \Omega_S(\alpha, \beta)$ for $|\lambda|$ sufficiently small. Let us suppose further that ψ and $\partial\psi/\partial t$ have partial derivatives with respect to λ;

$$\frac{\partial\psi}{\partial\lambda}(t, \lambda) \quad \text{and} \quad \frac{\partial}{\partial\lambda}\frac{\partial\psi}{\partial t}(t, \lambda)$$

which are continuous functions of (t, λ). Then

$$f(\varphi_\lambda) = \int_a^b F\left(t, \psi(t, \lambda), \frac{\partial\psi}{\partial t}(t, \lambda)\right) dt$$

is a differentiable function of the variable λ, and the lemma concerning differentiation under the integral sign says that its derivative with respect to λ, for $\lambda = 0$, is equal to

$$(2.7.4) \quad \int_a^b \left[\frac{\partial F}{\partial x}(t, \varphi_0(t), \varphi_0'(t)) \cdot \frac{\partial\psi}{\partial\lambda}(t, 0) + \frac{\partial F}{\partial y}(t, \varphi_0(t), \varphi_0'(t)) \cdot \frac{\partial}{\partial\lambda}\frac{\partial\psi}{\partial t}(t, 0)\right] dt.$$

If φ_0 gives a relative minimum of $f(\varphi)$, the derivative of $f(\varphi)$ with respect to λ necessarily vanishes for $\lambda = 0$; therefore the expression (2.7.4) must vanish. This said, it is evident that Theorem 2.7.1 will be proved provided we can establish the lemma:

Lemma 2.7.2. If $u: I \to (\mathbf{R})^n$ is a function of class C^1 satisfying the condition P(S), there exists $\psi(t, \lambda)$ as above, with values in S, such that

$$(2.7.5) \qquad \psi(t, 0) = \varphi_0(t), \qquad \frac{\partial\psi}{\partial t}(t, 0) = u(t), \qquad \frac{\partial}{\partial\lambda}\frac{\partial\psi}{\partial t}(t, 0) = u'(t).$$

(Intuitively, this means that the curve φ belongs to a family of curves φ_λ traced on S, in such a manner that for each value of $t \in I$, the "infinitesimal displacement" $\dot{\psi}(t, \lambda)$ which occurs when λ varies about 0 is equal to the vector $u(t)$, tangential to S at the point $\varphi_0(t)$.)

We shall limit ourselves to a brief indication of the proof of the lemma, omitting the details. We shall suppose that the variety S is of class C^3 so that the ds^2 of this variety will be a function of class C^2 of the parameters which locally define S. Then the differential equation of the geodesics of S will be of class C^1, and we may apply general theorems concerning existence and dependence on initial conditions. With each point $x \in S$, and with each tangent vector $y \in T_x(S)$, we associate the geodesic

$$\lambda \to g(x, y, \lambda),$$

which is described by a parameter proportional to its arc length, and satisfies $g(x, y, 0) = x$ and $(\partial g/\partial\lambda)(x, y, 0) = y$; it is defined if $\|\lambda y\|$ is sufficiently small. The function $g(x, y, \lambda)$ so defined is of class C^1. Replace x and y in $g(x, y, \lambda)$ by $\varphi_0(t)$ and $u(t)$ respectively; $g(\varphi_0(t), u(t), \lambda)$ is a function $\psi(t, u)$ satisfying the conditions of the lemma.

DEFINITION. A curve $\varphi_0: I \to S$ which satisfies the necessary condition of Theorem 2.7.1 is called an *extremal* of $\int_a^b F(t, \varphi(t), \varphi'(t))\, dt$ traced on S.

Writing φ instead of φ_0, we see that *a curve* $\varphi\colon \mathrm{I} \to \mathrm{S}$ *is an extremal if and only if*

$$\int_a^b \left[\frac{\partial \mathrm{F}}{\partial x}(t, \varphi(t), \varphi'(t))\cdot u(t) + \frac{\partial \mathrm{F}}{\partial y}(t, \varphi(t), \varphi'(t))\cdot u'(t)\right] dt = 0 \tag{2.7.6}$$

for every function $u\colon \mathrm{I} \to \mathbf{R}^n$ *of class* C^1 *satisfying the condition* P(S).

2.8 *Transformation of the preceding condition*

We shall proceed as in § 1.5. But, for simplicity, *we shall suppose* that the function of t

$$\frac{\partial \mathrm{F}}{\partial y}(t, \varphi(t), \varphi'(t))$$

has a derivative when φ is an extremal. As in § 1.5, put

$$\mathrm{A}(t) = \frac{\partial \mathrm{F}}{\partial x}(t, \varphi, \varphi'(t)), \qquad \mathrm{B}(t) = \frac{\partial \mathrm{F}}{\partial y}(t, \varphi(t), \varphi'(t));$$

an integration by parts gives

$$\int_a^b (\mathrm{A}(t)\cdot u(t) + \mathrm{B}(t)\cdot u'(t))\, dt = \int_a^b (\mathrm{A}(t) - \mathrm{B}'(t))\cdot u(t)\, dt,$$

since $u(a) = 0$ and $u(b) = 0$. Thus the condition for φ to be an extremal becomes

$$\int_a^b (\mathrm{A}(t) - \mathrm{B}'(t))\cdot u(t)\, dt = 0$$

for every function $u\colon \mathrm{I} \to \mathbf{R}^n$ of class C^1 satisfying condition P(S). By reasoning similar to that used in the proof of Lemma 1.5.3, it may be shown that, for every $t \in \mathrm{I}$, the linear form $\mathrm{A}(t) - \mathrm{B}'(t) \in \mathscr{L}(\mathbf{R}^n; \mathbf{R})$ *must vanish on every tangential vector to* S *at the point* $\varphi(t)$. This is left as an exercise for the reader. Of course, this necessary condition is also sufficient. In conclusion:

THEOREM 2.8.1. *In order that* $\varphi\colon \mathrm{I} \to \mathrm{S}$ *be an extremal, and with the proviso that*

$$\frac{\partial \mathrm{F}}{\partial y}(t, \varphi(t), \varphi'(t))$$

is a differentiable function of t, *it is necessary and sufficient that, for every* $t \in \mathrm{I}$, *the element*

$$\frac{\partial \mathrm{F}}{\partial x}(t, \varphi(t), \varphi'(t)) - \frac{d}{dt}\frac{\partial \mathrm{F}}{\partial y}(t, \varphi(t), \varphi'(t)) \in \mathscr{L}(\mathbf{R}^n; \mathbf{R})$$

be zero on all the tangent vectors of the space $\mathrm{T}_x(\mathrm{S})$ *at the point* $x = \varphi(t)$.

In this case, this condition replaces the Euler equation of Theorem 1.5.2.

Example. Take $n = 3$, and let $\mathrm{S} \subset \mathbf{R}^3$ be a *surface* in 3-dimensional Euclidean space. Take

$$\mathrm{F}(t, x_1, x_2, x_3, y_1, y_2, y_3) = \sqrt{(y_1)^2 + (y_2)^2 + (y_3)^2}.$$

The integral $\int F(t, \varphi(t), \varphi'(t))\, dt$ is thus the arc length, and the extremals are the *geodesics* of the surface S; they may be considered as curves in the ambient space $\mathbf{R}^3$. Let us interpret the condition of Theorem 2.8.1: the linear form $\partial F/\partial y$, of coefficients $\partial F/\partial y_1$, $\partial F/\partial y_2$, $\partial F/\partial y_3$ is that which has the coefficients

$$\frac{\partial F}{\partial y_i} = \frac{y_i}{\sqrt{y_1^2 + y_2^2 + y_3^2}}$$

which are the *direction cosines of the tangent* to the curve φ. Therefore the $(d/dt)(\partial F/\partial y_i)$ are the components of the derivative (with respect to t) of the unit tangent vector to the curve φ; they are proportional to the components of the unit vector of the *principal normal* to the curve φ. Thus, Theorem 2.8.1 states that the principal normal at the point $\varphi(t)$ is orthogonal to all the tangent vectors of S at the point $\varphi(t)$. In other words: *the geodesics of* S *are the curves such that the principal normal to the curve is normal to the surface* S *at each point of the curve.*

3. Problems in two dimensions

So far we have confined ourselves to the "variation" of a simple integral $\int_a^b F(t, \varphi(t), \varphi'(t))\, dt$ along a curve $\varphi: I \to E$. We now give a brief discussion of the variation of a *double integral.*

3.1 *Introduction of the problem*

Instead of a segment $I \in \mathbf{R}$, we shall consider a *compact set* with boundary $K \subset \mathbf{R}^2$ (with coordinates t_1, t_2 in the $\mathbf{R}^2$-plane). We say that a mapping

$$\varphi: K \to E \qquad \text{(where E is a Banach space)}$$

is of class C^1 if it is the restriction to K of a mapping of class C^1 of an open neighbourhood of K. The mappings $\varphi: K \to E$ of class C^1 form a vector space V. We have the following norm on V:

$$\|\varphi\| = \sup_K \|\varphi(t_1, t_2)\| + \sup_K \left\| \frac{\partial \varphi}{\partial t_1}(t_1, t_2) \right\| + \sup_K \left\| \frac{\partial \varphi}{\partial t_2}(t_1, t_2) \right\|$$

It may be verified that V is *complete* with respect to this norm (the proof has tedious complications, and will not be given here); V is therefore a Banach space.

Suppose now that a numerical function $F(t_1, t_2, x, y_1, y_2)$ is given in an open $U \subset \mathbf{R} \times \mathbf{R} \times E \times E \times E$, and is of class C^k $(k \geqslant 1)$.

Those mappings $\varphi \in V$ satisfying

$$\left[t_1, t_2, \varphi(t_1, t_2), \frac{\partial \varphi}{\partial t_1}(t_1, t_2), \frac{\partial \varphi}{\partial t_2}(t_1, t_2)\right] \in U$$

for all $(t_1, t_2) \in K$ constitute an *open* subset Ω of the Banach space V (the proof is analogous to that of Prop. 1.2.1; it involves use of the compactness of K). For $\varphi \in \Omega$, we define the "functional"

$$f(\varphi) = \iint_K F\left[t_1, t_2, \varphi(t_1, t_2), \frac{\partial\varphi}{\partial t_1}(t_1, t_2), \frac{\partial\varphi}{\partial t_2}(t_1, t_2)\right] dt_1 \wedge dt_2.$$

We show that $f\colon \Omega \to \mathbf{R}$ is of class C^k (as is F), and that, for every vector $u \in V$, it is true that

$$(3.1.1)\quad f'(\varphi)u = \iint_K \left\{\frac{\partial F}{\partial x}\cdot u(t_1, t_2) + \frac{\partial F}{\partial y_1}\cdot\frac{\partial u}{\partial t_1}(t_1, t_2) + \frac{\partial F}{\partial y_2}\cdot\frac{\partial u}{\partial t_2}(t_1, t_2)\right\} dt_1 \wedge dt_2$$

where, in

$$\frac{\partial F}{\partial x}, \quad \frac{\partial F}{\partial y_1} \quad \text{and} \quad \frac{\partial F}{\partial y_2},$$

x is equal to $\varphi(t_1, t_2)$, y_1 to $\partial u/\partial t_1$, and y_2 to $\partial u/\partial t_2$.

In § 1.4 the minimum problem for φ, such that $\varphi(a)$ and $\varphi(b)$ take prescribed values, was posed. Here the extremities a and b of the segment I are replaced by the *boundary* ∂K of our compact set K. We shall therefore limit ourselves to those $\varphi\colon K \to E$ such that the restriction of φ to ∂K is equal to a *given* mapping $\psi\colon \partial K \to E$ of class C^1. These φ constitute an affine subspace $W(\psi)$ of the Banach space V; the given mapping of a $\varphi_0 \in W(\psi)$ defines a bijection of $W(\psi)$ onto $W(0)$, viz.,

$$\varphi \to \varphi - \varphi_0$$

(the function $\varphi - \varphi_0$ vanishes on the boundary ∂K).

Given a $\varphi_0 \in \Omega$ we may enquire if it gives a *minimum* of $f(\varphi)$ among those φ sufficiently close to φ_0 which have *the same restriction to* ∂K *as* φ_0. A necessary condition for a minimum is that the derivative $f'(\varphi_0)$ vanish on all vectors u of the vector subspace $W(0)$. By definition, such a φ_0 will be called an *extremal* for the integral

$$\iint_K F\left(t_1, t_2, \varphi(t_1, t_2), \frac{\partial\varphi}{\partial t_1}, \frac{\partial\varphi}{\partial t_2}\right) dt_1 \wedge dt_2.$$

Thus, writing φ instead of φ_0, and taking account of equation (3.1.1), we have:

THEOREM 3.1.1. *In order that* $\varphi\colon K \to E$, *in* Ω, *be an extremal, it is necessary and sufficient that*

$$(3.1.2)\quad \iint_K \left\{\frac{\partial F}{\partial x}\cdot u(t_1, t_2) + \frac{\partial F}{\partial y_1}\cdot\frac{\partial u}{\partial t_1}(t_1, t_2) + \frac{\partial F}{\partial y_2}\,\frac{\partial u}{\partial t_2}(t_1, t_2)\right\} dt_1 \wedge dt_2 = 0$$

for arbitrary $u\colon K \to E$ *of class* C^1, *vanishing everywhere on the boundary* ∂K. [On the left-hand side of (3.1.2), $\partial F/\partial x$ denotes:

$$\frac{\partial F}{\partial x}\left(t_1, t_2, \varphi(t_1, t_2), \frac{\partial\varphi}{\partial t_1}, \frac{\partial\varphi}{\partial t_2}\right), \quad \text{and similarly for } \frac{\partial F}{\partial y_1} \text{ and } \frac{\partial F}{\partial y_2}.\Bigg]$$

This theorem is analogous to Theorem 1.4.1.

3.2 *Transformation of the condition for an extremum*

Set

$$(3.2.1)\qquad \begin{cases} \dfrac{\partial F}{\partial x}\left(t_1, t_2, \varphi(t_1, t_2), \dfrac{\partial\varphi}{\partial t_1}, \dfrac{\partial\varphi}{\partial t_2}\right) = A(t_1, t_2) \\ \dfrac{\partial F}{\partial y_1}\left(t_1, t_2, \varphi(t_1, t_2), \dfrac{\partial\varphi}{\partial t_1}, \dfrac{\partial\varphi}{\partial t_2}\right) = B_1(t_1, t_2) \\ \dfrac{\partial F}{\partial y_2}\left(t_1, t_2, \varphi(t_1, t_2), \dfrac{\partial\varphi}{\partial t_1}, \dfrac{\partial\varphi}{\partial t_2}\right) = B_2(t_1, t_2). \end{cases}$$

These are continuous functions on K, with values in $\mathscr{L}(E; \mathbf{R})$.

Further, we shall suppose that B_1 and B_2 are of class C^1; and under this restrictive hypothesis we shall prove the lemma:

Lemma 3.2.1. In order that the integral

$$(3.2.2)\qquad \iint_K \left(A(t_1, t_2)\cdot u(t_1, t_2) + B_1(t_1, t_2)\cdot\frac{\partial u}{\partial t_1} + B_2(t_1, t_2)\frac{\partial u}{\partial t_2}\right) dt_1 \wedge dt_2$$

vanish for every $u\colon K \to E$ of class C^1, which vanishes on ∂K, it is necessary and sufficient that

$$(3.2.3)\qquad A = \frac{\partial B_1}{\partial t_1} + \frac{\partial B_2}{\partial t_2}.$$

ABRIDGED PROOF. The function $u\colon K \to E$ being given, let us consider the differential form of degree 1, of two variables t_1 and t_2:

$$\omega = [B_1(t_1, t_2)\cdot u(t_1, t_2)]\, dt_2 - [B_2(t_1, t_2)\cdot u(t_1, t_2)]\, dt_1.$$

We have

$$\left(B_1(t_1, t_2)\frac{\partial u}{\partial t_1} + B_2(t_1, t_2)\frac{\partial u}{\partial t_2}\right) dt_1 \wedge dt_2 = d\omega - \left[\left(\frac{\partial B_1}{\partial t_1} + \frac{\partial B_2}{\partial t_2}\right)\cdot u(t_1, t_2)\right] dt_1 \wedge dt_2;$$

therefore the integral (3.2.2) is equal to

$$\iint_K \left(A - \frac{\partial B_1}{\partial t_1} - \frac{\partial B_2}{\partial t_2}\right)\cdot u(t_1, t_2)\, dt_1 \wedge dt_2 + \iint_K d\omega.$$

Now, by Stokes's theorem

$$\iint_K d\omega = \int_{\partial K} \omega;$$

and this vanishes since ω is zero on ∂K, $u(t_1, t_2)$ being 0 for every $(t_1, t_2) \in \partial K$.

Thus, for every u of class C^1, vanishing on ∂K, the integral (3.2.2) is equal to

$$\iint_K \left(A - \frac{\partial B_1}{\partial t_1} - \frac{\partial B_2}{\partial t_2}\right)\cdot u(t_1, t_2)\, dt_1 \wedge dt_2.$$

In order that this integral vanish for arbitrary u (zero on ∂K), it is necessary (and evidently sufficient) that the function $A - \partial B_1/\partial t_1 - \partial B_2/\partial t_2$ be *identically zero*: the reasoning is analogous to that used in the proof of Lemma 1.5.3, and is left to the reader. Thus the condition (3.2.3) is necessary and sufficient.

Q.E.D.

Taking account of the values of A, B_1, and B_2 given by (3.2.1), we have proved:

THEOREM 3.2.2. *In order that* $\varphi\colon K \to E$, *belonging to* Ω, *be an extremal, it is necessary and sufficient that*

$$\boxed{\frac{\partial F}{\partial x} = \frac{\partial}{\partial t_1}\left(\frac{\partial F}{\partial y_1}\right) + \frac{\partial}{\partial t_2}\left(\frac{\partial F}{\partial y_2}\right)} \tag{3.2.4}$$

it being understood that in $\partial F/\partial x$, $\partial F/\partial y$, and $\partial F/\partial y_2$, x is equal to $\varphi(t_1, t_2)$, y_1 to $\partial\varphi/\partial t_1$ and y_2 to $\partial\varphi/\partial t_2$. This equation replaces the Euler equation.

Remark. This equation has only been established under the assumption that $\partial F/\partial y_1$ and $\partial F/\partial y_2$ are of class C^1 in t_1 and t_2.

Relation (3.2.4) states that

$$\boxed{\frac{\partial F}{\partial x}\,dt_1 \wedge dt_2 = d\left(\frac{\partial F}{\partial y_1}\,dt_2 - \frac{\partial F}{\partial y_2}\,dt_1\right)}\,; \tag{3.2.5}$$

this is an alternative form of the equation of extremals.

When the function $F(t_1, t_2, x, y_1, y_2)$ is independent of x, the condition (3.2.5) expresses simply this: the differential form

$$\frac{\partial F}{\partial y_1}\left(t_1, t_2, \frac{\partial\varphi}{\partial t_1}, \frac{\partial\varphi}{\partial t_1}\right) dt_2 - \frac{\partial F}{\partial y_2}\left(t_1, t_2, \frac{\partial\varphi}{\partial t_1}, \frac{\partial\varphi}{\partial t_2}\right) dt_1$$

is *closed* (in other words its exterior differential vanishes identically).

Example. Let x, y, z be coordinates in the space $\mathbf{R}^3$, and consider the surfaces defined by $z = \varphi(x, y)$. In the preceding theory, we take $E = \mathbf{R}$, and replace t_1 by x, t_2 by y, x by z, $\partial\varphi/\partial t_1$ by $\partial z/\partial x$ and $\partial\varphi/\partial t_2$ by $\partial z/\partial y$; we shall use the usual notation: $p = \partial z/\partial x$, $q = \partial z/\partial y$.

Suppose we are given a function $F(p, q)$, independent of x, y, z; consider the functional

$$\iint_K F(p, q)\,dx \wedge dy$$

taken over a part of the surface $z = \varphi(x, y)$, with $(x, y) \in K$. The extremal surfaces are those for which the differential form

$$\frac{\partial F}{\partial p}\,dy - \frac{\partial F}{\partial q}\,dx$$

is a *closed form.*

More particularly, consider

$$F(p, q) = \sqrt{1 + p^2 + q^2}; \quad \text{then} \quad \sqrt{1 + p^2 + q^2}\,dx \wedge dy$$

is the *area element* of the surface $z = \varphi(x, y)$. The extremal surfaces for the integral

$$\iint \sqrt{1 + p^2 + q^2}\, dx \wedge dy$$

are called *minima surfaces* (because a sufficiently small part of such a surface realises a minimum of the area *vis-à-vis* "neighbouring" surfaces having the same boundary). By (3.2.4) the minima surfaces are the solutions of

$$\frac{\partial}{\partial x}\left(\frac{p}{\sqrt{1 + p^2 + q^2}}\right) + \frac{\partial}{\partial y}\left(\frac{q}{\sqrt{1 + p^2 + q^2}}\right) = 0. \tag{3.2.6}$$

This is a partial differential equation of the second order, for the unknown function z of the two variables x and y. It reduces to

$$(1 + q^2)r - 2pqs + (1 + p^2)t = 0, \tag{3.2.7}$$

where r, s, t denote $\partial^2 z/\partial x^2$, $\partial^2 z/\partial x \partial y$, $\partial^2 z/\partial y^2$ respectively.

Exercises

1. Define a norm in the space of curves of class C^2 on $[a, b]$ (with values in a Banach space E) by

$$\|x\| = \sup_{t \in [a, b]} (\|x(t)\| + \|x'(t)\| + \|x''(t)\|).$$

If $F(t, x, y, z)$ is a function of class C^1, put

$$f(x) = \int_a^b F(t, x(t), x'(t), x''(t))\, dt.$$

Show that f is a differentiable function of x with derivative

$$f'(x) \cdot u = \int_a^b \left(\frac{\partial F}{\partial x} \cdot u + \frac{\partial F}{\partial y} \cdot u' + \frac{\partial F}{\partial z} \cdot u''\right) dt$$

(where, for simplicity, we have written

$$\frac{\partial F}{\partial x}(t, x(t), x'(t), x''(t)) \cdot u(t) = \frac{\partial F}{\partial x} \cdot u, \text{ etc.} \ldots).$$

Transform this derivative by integration by parts.

2. Determine the extremals of the integrals

$$\int_a^b t x' y'\, dt, \qquad \int_a^b (x' + y)(x + y')\, dt, \qquad \int_a^b x(x' + y' + t)\, dt.$$

Describe in general terms the nature of the extremals of

$$I(f) = \int F(t, f, f')\, dt$$

if F is linear or affine in the third variable.

3. Consider an integral of the form

$$\mathrm{I}(x) = \int_a^b \mathrm{F}(t, x') \, dt$$

where F is a numerical function of class C^1 in the plane. Let $x_0(t)$ be an extremal, if it exists, such that $x_0(a) = \alpha$, $x_0(b) = \beta$, and let $x(t)$ be a second extremal of class C^1 such that $x(a) = \alpha$, $x(b) = \beta$.

Show that

$$\mathrm{I}(x) - \mathrm{I}(x_0) = \int_a^b \left[\mathrm{F}(t, x') - \mathrm{F}(t, x_0') - (x' - x_0') \frac{\partial \mathrm{F}}{\partial x'} (t, x_0'(t))\right] dt.$$

Deduce that if F is a convex function of its second variable, then x_0 gives an absolute minimum (use Exercise 8 of Chap. 1 of *Differential Calculus*).

4. Study the existence and uniqueness of the extremals of the integral

$$\int_a^b (x^2 - x'^2) \, dt$$

which satisfy the conditions $x(a) = \alpha$ and $x(b) = \beta$.

5. Suppose that λ is a known constant, and consider the extremals of the integral

$$\mathrm{I}_\lambda(x) = \int_{-1}^{+1} (t^2 + \lambda^2) x'^2 \, dt.$$

(a) Show that if $\lambda \neq 0$ there exists a unique extremal which passes through the points $(-1, a)$ and $(1, b)$ $(a \neq b)$, and that it gives an absolute minimum for the integral $\mathrm{I}_\lambda(x)$.
(b) If $\lambda = 0$, show that there exist no extremals of class C^1. Show that $\mathrm{I}_0(x)$ is never zero, but that it can be made as small as desired by a suitable choice of the function $x(t)$.

6. (a) *Preliminaries.* Let Ω be an open subset of a Banach space E, f and g real functions of class C^1 in Ω.

C being a given constant, consider the set $V \subset \Omega$ of those points x satisfying $g(x) = C$. Let a be a point of V; suppose that $g'(a) \neq 0$. Show that the codimension of the kernel N of $g'(a)$ is 1.

Suppose $u \in N$, $u \neq 0$; show that there exists an element of a variety of dimension 1, of class C^1, contained in V, passing through a, and tangential to a vector u at a. (Find the intersection of V with the 2-dimensional plane defined by $x = a + \lambda u + \mu v$, where v is fixed and $v \notin N$, and show, by means of the implicit function theorem, that the restriction of this intersection to a neighbourhood of a is the required variety element.)

Deduce that if the restriction of f to V admits of an extremum at a, then $f'(a) \cdot u = 0$ for every $u \in N$; next show that $f'(a)$ is proportional to $g'(a)$ (it can be shown that

$$f'(a) - \frac{f'(a) \cdot v}{g'(a) \cdot v} g'(a) = 0).$$

(b) *Application.* Let I be the interval $[a, b]$; $\mathrm{F}(t, x, y)$, $\mathrm{G}(t, x, y)$ two functions of class C^1 in $\mathrm{I} \times \mathbf{R}^n \times \mathbf{R}^n$; α, β, C three given numbers.

Put

$$f(x) = \int_a^b \mathrm{F}(t, x(t), x'(t)) \, dt$$

$$g(x) = \int_a^b \mathrm{G}(t, x(t), x'(t)) \, dt.$$

Consider the functions $x(t)$, of class C^1 on $[a, b]$, which satisfy

$$\begin{aligned} x(a) &= \alpha \\ x(b) &= \beta \\ y(x) &= C, \end{aligned}$$

and make $f(x)$ an extremum.

Show that if such a function $x_0(t)$ exists and if $g'(x_0) \neq 0$, then there exists a number λ such that $x_0(t)$ is an extremal $y_0(t, \lambda)$ of the integral $f(x) - \lambda g(x)$. The number λ is determined *a posteriori* by the condition

$$g(y_0(\lambda)) = C.$$

Example. Determine the curves Γ in $\mathbf{R}^2$ which have minimal lengths and which join the origin to the point (1, 0) under the condition that $\int_\Gamma P\,dx + Q\,dy = C$, where (P, Q) is a vector field of class C^1 in $\mathbf{R}^2$. (Parametrize Γ by its arc length s, and obtain the system of second order differential equations satisfied by $x(s)$ and $y(s)$.)

7. Using the first part of the preceding exercise, determine those functions of class C^1 which make the integral

$$\int_0^{2\pi} (r^2 + r'^2)^{\frac{1}{2}}\, du$$

a minimum, and which satisfy

$$\int_0^{2\pi} r^2\, du = C,\ r(0) = r(2\pi) = 0$$

(C a given positive constant).

[This determines, in polar coordinates, those curves enclosing a given area and having a minimal length.]

8. Let $\Sigma\,(f)$ denote the surface of revolution defined, in cylindrical polar coordinates, (r, θ, z), by $r = f(z)$, $a \leqslant z \leqslant b$ (f being a positive function of class C^1).

(a) Write the area $A(f)$ of this surface in the form of a simple integral over $[a, b]$, and determine the functions f which are extremals of this integral (first find the inverse function of f).

(b) For $a \leqslant z_1 \leqslant z_2 \leqslant b$, denote by $V(z_1, z_2, f)$ the volume of the domain of $\mathbf{R}^3$ defined by the inequalities $0 \leqslant r \leqslant f(z)$, $z_1 \leqslant z \leqslant z_2$, and by $A(z_1, z_2, f)$ the area of that part of $\Sigma\,(f)$ lying between the planes $z = z_1$, $z = z_2$.

Show that the extremal surfaces determined in (a), together with the cylinders of revolution, are the only surfaces of revolution about $0z$ for which the ratio

$$\frac{V(z_1, z_2, f)}{A(z_1, z_2, f)}$$

is independent of z_1 and z_2.

9. Let $f(t, x, y)$ be a numerical function of class C^2 defined in $\Omega \times \mathbf{R}$, Ω being an open convex subset of the plane; consider the functions f which satisfy the condition

$$\text{(D)} \qquad \frac{\partial f}{\partial x} - \frac{\partial^2 f}{\partial t \partial y} - y\frac{\partial^2 f}{\partial x \partial y} = 0.$$

If A, B denote two points of Ω with abscissae a, b, $a \neq b$, consider

$$\int_a^b f(t, x, x')\, dt.$$

(a) Show that condition (D) is satisfied if all straight line segments in Ω are extremals.
(b) Let $X(t)$ denote the linear function whose graph joins A and B. Interpret condition (D) when f depends linearly on y; show that

$$\int_a^b f(t, x, x')\, dt = \int_a^b f(t, X, X')\, dt$$

for every function $x(t)$ of class C^2 whose graph joins A and B in Ω.
(c) Show that if f is a convex function of y, then

$$\int_a^b f(t, x, x')\, dt \geqslant \int_a^b f(t, X, X')\, dt$$

in the notation of the preceding question.

10. Prove the two following propositions by means of Schwarz's inequality (analogous respectively to Prop. 2.5.1 and 2.5.2):
(1) If a curve parametrized by $x = \varphi(t)$ gives a *relative minimum* for

$$\int_a^b F(\varphi(t), \varphi'(t))\, dt$$

(*vis-à-vis* those curves of class C^1 with parameter $t \in [a, b]$, and sufficiently close to x in the sense of the topology of the space V), then it also gives a relative minimum for

$$\int_a^b \sqrt{F(\varphi(t), \varphi'(t))}\, dt.$$

(2) If a geometrical curve gives a *relative minimum* for

$$\int_a^b \sqrt{F(\varphi(t), \varphi'(t))}\, dt$$

and if a parametrization is adopted which makes $F(\varphi(t), \varphi'(t))$ constant (cf. Prop. 2.5.2), then it gives a relative minimum for

$$\int_a^b F(\varphi(t), \varphi'(t))\, dt.$$

11. Instead of supposing, as in § 2.5, that $F(x, y)$ is homogeneous-quadratic in y, suppose more generally that

$$F = F_2 + F_0,$$

where $F_2(x, y)$ is a non-degenerate, homogeneous quadratic in y, and F_0 is independent of y. By Theorem 2.4.1, $F_2 - F_0$ is constant along each extremal of $\int F\, dt$. Consider an extremal on which

$$F_2 - F_0 = h,$$

h being a given constant, and suppose that $\varphi'(t) \neq 0$ for every t. Then $(F_0 + h)F_2 > 0$ at every point of this extremal, since $F_2(x, y) \neq 0$ for $y \neq 0$. For every sufficiently close curve (in the sense of the topology of V), it follows that $(F_0 + h)F_2 > 0$. Let U be the open set formed by the points (x, y) such that

$$(F_0(x) + h)F_2(x, y) > 0;$$

and consider the problem of the calculus of variations for the integral

$$\int_a^b \sqrt{(F_0 + h)F_2}\, dt.$$

Prove the two following propositions which are analogous to Prop. 2.5.1, 2.5.2: every extremal of $\int F\,dt$ along which $F_2 - F_0 = h$, is also an extremal of

$$\int \sqrt{(F_0 + h)F_2}\,dt;$$

conversely, if a geometrical curve is an extremal of

$$\int \sqrt{(F_0 + h)F_2}\,dt,$$

there exists a parametrization for this curve such that $F_2 = F_0 + h$, and, with this parametrization, this curve is an extremal of $\int F\,dt$.

12. Determine the geodesics of the surface defined parametrically by

$$\begin{aligned} x &= \tanh u \cos v \\ y &= \tanh u \sin v \\ z &= \frac{1}{\cosh u} + \log \tanh (u/2). \end{aligned}$$

13. In the notation of § 2.6, consider the explicit representation

$$F(u, u') = \sum_{i,j} g_{ij}(u) u_i' u_j', \qquad (g_{ij} = g_{ji}).$$

Verify that (2.6.4) becomes

$$\sum_j g_{ij}(u) u_j'' = \tfrac{1}{2} \sum_{j,k} \left(\frac{\partial g_{jk}}{\partial u_i} - \frac{\partial g_{ij}}{\partial u_k} - \frac{\partial g_{ik}}{\partial u_j} \right) u_j' u_k'.$$

If $(g^{ij}(u))$ denotes the matrix inverse to $(g_{ij}(u))$, equation (2.6.5) becomes

$$\boxed{u_i'' + \sum_{j,k} \Gamma^i_{j,k}(u) u_j' u_k' = 0},$$

where the "Riemann–Christoffel symbol" $\Gamma^i_{j,k}$ is defined by

$$\Gamma^i_{j,k} = \tfrac{1}{2} \sum_h g^{ih} \left(-\frac{\partial g_{jk}}{\partial u_h} + \frac{\partial g_{hj}}{\partial u_k} + \frac{\partial g_{hk}}{\partial u_j} \right).$$

14. Write down the Riemann–Christoffel symbols (see preceding exercise) for the quadratic form

$$\varphi(x, y, x', y') = E(x, y)x'^2 + 2F(x, y)x'y' + G(x, y)y'^2.$$

(a) Show that the necessary and sufficient condition which must be satisfied if the curves $y = \text{const.}$ are geodesics is

$$F \frac{\partial E}{\partial x} + E \frac{\partial E}{\partial y} = 2G \frac{\partial F}{\partial x}.$$

(b) If, further, $F = 0$, show that φ can be written in the form $u'^2 + H(u, v)v'^2$.

Example. Determine the geodesics when $\varphi = u'^2 + e^{2u}v'^2$.

15. Let

$$F(a, b, c, d, e) = -\frac{a^2}{d^3e^3} + \frac{bc}{d^2e^2},$$

a function of five real variables defined for $de \neq 0$.

Consider an open D of **R** such that $\overline{D}$ is a boundary compact set of the plane which does not meet the coordinate axes.

(a) Let f be a function of class C^2 in a neighbourhood of $\overline{D}$. Show that a necessary condition for f to be an extremal of the integral

$$I(f) = \iint_D F(f, f'_{t_1}, f'_{t_2}, t_1, t_2)\, dt_1\, dt_2$$

is that

$$f - t_1 f'_{t_1} - t_2 f'_{t_2} + t_1 t_2 f''_{t_1 t_2} = 0. \tag{1}$$

(b) Show that if f is of class C^3, the function $g(t_1, t_2) = f''_{(t_1)^2}(t_1, t_2)$ satisfies

$$g - t_2 g'_{t_2} = 0. \tag{2}$$

Integrate (2) and deduce that the general solution of (1) is

$$f(t_1, t_2) = t_1 \varphi(t_2) + t_2 \psi(t_1),$$

where φ and ψ are arbitrary functions.

(c) Determine the solutions of (1) which are such that on the line $t_2 = t_1$ it is true that

$$f(t_1, t_2) = t_2^2, \qquad f'_{t_1}(t_1, t_2) = t_2.$$

Chapter 3

Applications of the moving frame method to the theory of curves and surfaces

1. The moving frame

1.1 *Definition of the differential forms* ω_i *and* ω_{ij}

We shall be concerned with the space $\mathbf{R}^n$. A transformation T of the *linear-affine group* is determined when we are given: (1) the point $M \in \mathbf{R}^n$ which is the image of the origin $0 \in \mathbf{R}^n$; (2) the vectors $\vec{e}_1, \ldots, \vec{e}_n$ which are respectively the images of the vectors $(1, 0, \ldots, 0), \ldots, (0, \ldots, 0, 1)$ of the canonical base of $\mathbf{R}^n$ under the linear homogeneous transformation associated with T.

The only condition which must be imposed is:

$$\det(\vec{e}_1, \ldots, \vec{e}_n) \neq 0, \tag{1.1.1}$$

The data

$$(M, \vec{e}_1, \ldots, \vec{e}_n)$$

satisfying (1.1.1) is called an *affine frame*; the point M is the origin of the frame.

If we abandon condition (1.1.1), the set $(M, \vec{e}_1, \ldots, \vec{e}_n)$ is identified with the product space

$$\underbrace{\mathbf{R}^n \times \cdots \times \mathbf{R}^n}_{(n+1)\text{-times}} = \mathbf{R}^{n(n+1)}$$

since $M, \vec{e}_1, \ldots, \vec{e}_n$ take their values in $\mathbf{R}^n$. The condition (1.1.1) means that this point of $\mathbf{R}^{n(n+1)}$ does not belong to the *closed* set defined by

$$\det(\vec{e}_1, \ldots, \vec{e}_n) = 0.$$

Thus, the *affine frames constitute an open set* $U \subset \mathbf{R}^{n(n+1)}$; and $M, \vec{e}_1, \ldots, \vec{e}_n$ are functions $U \to \mathbf{R}^n$ (induced by the $(n+1)$ projections of $\mathbf{R}^{n(n+1)}$ onto its $(n+1)$ factors). These functions are evidently differentiable.

We can consider frames $r \in U$ which depend on one or more real parameters. A particular case is that in which the parameters are taken to be the coordinates of the point $U \subset \mathbf{R}^{n(n+1)}$; in other words, this is the case where r is considered as a function of itself, i.e., of the point $r \in U$. In general, a parameter dependent frame r is called a "*moving frame*"; we shall consider the rudiments of this theory.

When we have a differentiable function f, in an open U of a Banach space (here: $\mathbf{R}^{n(n+1)}$), with values in a Banach space E (here: with values in $\mathbf{R}^n$), we can always consider the *differential* df: this is a differential form of degree one, defined in U, with values in E (cf. Chap. 1, § 2.3). Thus, in the present case we have $(n + 1)$ differential forms in the open set U of frames, viz.,

$$dM, d\vec{e}_1, \ldots, d\vec{e}_n;$$

these are differential forms with values in $\mathbf{R}^n$. Each of these has n components with values in $\mathbf{R}$, which are differential forms with scalar values.

We now introduce a fundamental idea in the theory of the moving frame. The differential forms $dM, d\vec{e}_1, \ldots, d\vec{e}_n$ are functions ω of 2 variables r and $\xi (r \in U, \xi \in \mathbf{R}^{n(n+1)})$, linear in ξ for each $r \in U$, and with values in $\mathbf{R}^n$. But, for a given r, the vectors $\vec{e}_1(r), \ldots, \vec{e}_n(r)$ of the frame constitute a *base* of $\mathbf{R}^n$; therefore the value of $\omega(r; \xi)$ can be written uniquely as a linear combination $\sum_{i=1}^{n} a_i\vec{e}_i(r)$. The scalars a_i (for given r) are linear forms in ξ. Thus the functions $a_i(r, \xi)$ with values in $\mathbf{R}$ can be considered as *differential forms of degree one in* U, *with scalar values*. We can therefore write

$$dM = \sum_{i=1}^{n} \omega_i\vec{e}_i \tag{1.1.2}$$

$$d\vec{e}_i = \sum_{j=1}^{n} \omega_{ij}\vec{e}_j, \tag{1.1.3}$$

it being understood that for each $r \in U$, the notation $\vec{e}_i$ denotes the value $\vec{e}_i(r)$ of the ith vector of the frame r. We thus define $n(n + 1)$ *differential forms of degree one in* U, *with scalar values*, viz.:

$$\boxed{\omega_i \quad \text{and} \quad \omega_{ij}} \qquad (1 \leqslant i \leqslant n, 1 \leqslant j \leqslant n).$$

We can say that these forms define the "infinitesimal displacement" of the moving frame $r \in U$ as r varies.

1.2 *Relations satisfied by the forms* ω_i and ω_{ij}

We know that if f is a differentiable function (with vector values), then $d(df) = 0$ (cf. Chap. 1, Theorem 2.5.1). We shall write out in full the relations

$$d(dM) = 0, \qquad d(d\vec{e}_i) = 0.$$

For the first, take the differential d of the right-hand side of (1.1.2), taking account of the formula for the differential of a product. We have

$$d(\omega_i\vec{e}_i) = (d\omega_i)\vec{e}_i - \omega_i \wedge d\vec{e}_i,$$

because ω_i is a form of degree one. The relation $d(dM) = 0$ therefore gives, taking account of (1.1.3):

$$\sum_{j=1}^{n} (d\omega_j)\vec{e}_j = \sum_{i=1}^{n} \omega_i \wedge \left(\sum_{j=1}^{n} \omega_{ij}\vec{e}_j\right).$$

In this vectorial equality the coefficients of $\vec{e}_j$ on each side must be equal, since, for each frame r, the vectors $\vec{e}_j(r)$ form a base of $\mathbf{R}^n$. Hence,

$$(1.2.1) \qquad \boxed{d\omega_j = \sum_{i=1}^{n} \omega_i \wedge \omega_{ij}} \qquad 1 \leqslant j \leqslant n.$$

An analogous calculation for the relation $d(d\vec{e}_i) = 0$ gives:

$$(1.2.2) \qquad \boxed{d\omega_{ij} = \sum_{k=1}^{n} \omega_{ik} \wedge \omega_{kj}} \qquad 1 \leqslant i \leqslant n, 1 \leqslant j \leqslant n.$$

1.3 *Orthonormal frames*

Consider a frame $r = (\mathrm{M}, \vec{e}_1, \ldots, \vec{e}_n)$; in order that a linear-affine transformation T defined by r be a (Euclidean) *displacement*, it is necessary and sufficient that the linear homogeneous transformation associated with T be a transformation of the *orthogonal group* (group of those linear homogeneous transformations which preserve the scalar product). This means that it is necessary and sufficient that the base $(\vec{e}_1, \ldots, \vec{e}_n)$ be *orthonormal*, i.e., that the vectors $\vec{e}_1, \ldots, \vec{e}_n$ be of unit length (in the Euclidean sense) and mutually orthogonal. These conditions are expressed by:

$$(1.3.1) \qquad \vec{e}_i \cdot \vec{e}_j = \delta_{ij} \qquad (1 \text{ if } i = j, \quad 0 \text{ if } i \neq j),$$

where $\vec{a} \cdot \vec{b}$ denotes the scalar product of the two vectors $\vec{a}$ and $\vec{b}$. It may be shown (exercise) that the frames satisfying conditions (1.3.1) form a *variety* in the open U of all the affine frames. We shall denote this variety by O.

The differential forms ω_i and ω_{ij} "induce" (as we say) differential forms on the variety O. Let us be more precise: ω_i (for example) is a function $\omega_i(r, \xi)$ with values in $\mathbf{R}$, where $r \in \mathrm{U}$, $\xi \in \mathbf{R}^{n(n+1)}$; and, for fixed r, this is a linear function of ξ. The form induced by ω_i is the restriction of this function to $r \in O$ and to the ξ tangential to the variety O at the point r. (See Chap. 1, § 4.11.)

Of course, the forms induced on O by the forms ω_i and ω_{ij} still satisfy the relations (1.2.1) and (1.2.2). We shall continue to denote by ω_i and ω_{ij} the induced forms, since we shall only have need to consider the variety O of orthonormal frames. We see that on O the forms ω_i, ω_{ij} satisfy certain supplementary relations, obtained by differentiating (1.3.1): we have in fact

$$\vec{e}_i \cdot d\vec{e}_j + \vec{e}_j \cdot d\vec{e}_i = 0,$$

i.e., when $d\vec{e}_i$ and $d\vec{e}_j$ are replaced by their values from (1.1.3)

$$\vec{e}_i \cdot \Big(\sum_k \omega_{jk}\vec{e}_k\Big) + \vec{e}_j \cdot \Big(\sum_k \omega_{ik}\vec{e}_k\Big) = 0.$$

Since $\vec{e}_i \cdot \vec{e}_k = \delta_{ik}$, we have

$$(1.3.2) \qquad \boxed{\omega_{ji} + \omega_{ij} = 0} \qquad \text{for arbitrary } i \text{ and } j;$$

and in particular

(1.3.3) $\boxed{\omega_{ii} = 0}$ for every i.

These are the conditions satisfied by an "infinitesimal displacement" of an *orthonormal* frame.

It is well known that if $\vec{e}_1, \ldots, \vec{e}_n$ are n unit vectors of $\mathbf{R}^n$ which are mutually orthogonal, then

$$\det(\vec{e}_1, \ldots, \vec{e}_n) = \pm 1.$$

We may briefly recall the reason for this: in order that an $n \times n$ matrix A define a transformation of the orthogonal group $0(n)$, it is necessary and sufficient that the product of A with its transpose A′ be equal to the unit matrix. Since $\det(\mathrm{A}') = \det(\mathrm{A})$, it follows that $(\det(\mathrm{A}))^2 = 1$, so that $\det(\mathrm{A}) = \pm 1$. Those orthonormal frames for which $\det(\vec{e}_1, \ldots, \vec{e}_n) = +1$ are called *direct* frames; they define *direct displacements*. The set of direct frames will be denoted by SO; is one of the two connected components of the variety O of all orthonormal frames. In the following we shall be interested only in the variety SO, and in the differential forms ω_i and ω_{ij} induced on SO. We shall often use "frame" to mean "direct orthonormal frame".

1.4 *The Frenet frame of an oriented curve in* $\mathbf{R}^3$

By means of two examples, we shall study a curve traced on the variety SO of frames; in other words, we shall consider a family of direct orthonormal frames which are differentiable functions of a real parameter t. Then the forms ω_i and ω_{ij} become differentiable forms of the variable t; in this case equations (1.2.1) and (1.2.2) are of no interest, since every differential form of degree 2 of a variable t is identically zero.

The first example, which is the subject of the present section, associates with a differentiable curve C of $\mathbf{R}^3$ (by "differentiable" we mean of class C^k, for k sufficiently large) a parameter-dependent family of orthonormal frames of $\mathbf{R}^3$. The curve C is defined by the differentiable function $\mathrm{M}(t)$ of the real variable t, with values in $\mathbf{R}^3$. With each t we associate the following frame: its origin is the point $\mathrm{M}(t)$ of the curve C; the vector $\vec{e}_1(t)$ is the unit tangent vector to the curve C at the point with parameter t, the vector being oriented in the direction of increasing t. The vector $\vec{e}_2(t)$ will be defined in a moment, after which $\vec{e}_3(t)$ is determined from the conditions that the frame be orthonormal and direct.

We can already write

$$d\mathrm{M} = \omega_1 \vec{e}_1 \tag{1.4.1}$$

since the vectors $d\mathrm{M}/dt$ and $\vec{e}_1$ are parallel (we suppose that $d\mathrm{M}/dt \neq 0$ for every t); in other words, the forms ω_2 and ω_3 of (1.1.2) are zero. On the other hand, whatever be the choice of $\vec{e}_2$ as a differentiable function of t, we shall have

$$d\vec{e}_1 = \omega_{12}\vec{e}_2 - \omega_{31}\vec{e}_3 \tag{1.4.2}$$

(since $\omega_{11} = 0$, $\omega_{13} + \omega_{31} = 0$); thus $d\vec{e}_1/dt$ is a vector *orthogonal* to $\vec{e}_1$. We make the restrictive hypothesis that $d\vec{e}_1/dt \neq 0$; then we choose the unit vector $\vec{e}_2(t)$ to be parallel to $d\vec{e}_1/dt$ (this can be done in two ways; but once a choice has been made for a particular value of t, that for neighbouring values of t is fixed by continuity). We have thus associated with each value of t a direct orthonormal frame $(M, \vec{e}_1, \vec{e}_2, \vec{e}_3)$; it is called the *Frenet frame* of the curve C at the point t. This family of frames with parameter t is thus defined by the curve C in $\mathbf{R}^3$.

With the above choice of $\vec{e}_2$, relation (1.4.2) tells us that $\omega_{31} = 0$; thus the equations of motion of the frame are:

$$(1.4.3)\qquad d\vec{e}_1 = \omega_{12}\vec{e}_2, \qquad d\vec{e}_2 = -\omega_{12}\vec{e}_1 + \omega_{23}\vec{e}_3, \qquad d\vec{e}_3 = -\omega_{23}\vec{e}_2.$$

As for the form ω_1, it is written *a priori* in the form $a(t)\,dt$; by (1.4.2) we have

$$\frac{dM}{dt} = a(t)\vec{e}_1,$$

thus $a(t) > 0$ is equal to the length of the vector dM/dt. This signifies that $a(t)\,dt$ is equal to the differential ds of the *arc length* s of the curve C (the arc length varies in the same sense as t). Let us take the arc length s as the parameter for C (defined to within an arbitrary additive constant). Then the differential forms ω_{12} and ω_{23} become $b(s)\,ds$ and $c(s)\,ds$. Thus we have introduced two functions on the curve C; by definition, b is the *curvature* and c the *torsion*, denoted respectively by $1/\rho$ and $1/\tau$. With this notation, the equations of motion of the Frenet frame become

$$(1.4.4)\qquad \left\{\boxed{\begin{array}{c} \dfrac{dM}{ds} = \vec{e}_1, \\[2ex] \dfrac{d\vec{e}_1}{ds} = \dfrac{1}{\rho}\vec{e}_2, \qquad \dfrac{d\vec{e}_2}{ds} = -\dfrac{1}{\rho}\vec{e}_1 + \dfrac{1}{\tau}\vec{e}_3, \qquad \dfrac{de_3}{ds} = -\dfrac{1}{\tau}e_2 \end{array}}\right. .$$

These are the well known results.

1.5 *The Darboux frame of an oriented curve* C *traced on an oriented surface* S *in* $\mathbf{R}^3$

A differentiable curve C is "oriented", at each point $M \in C$, by the choice of a unit tangent vector to C at the point M (this choice varying continuously with M). To "orientate" a surface S in $\mathbf{R}^3$ (supposed connected, and of class C^k for k sufficiently large), we must, in line with the definition already given (see Chap. 1, § 4.8), orientate the tangent plane to S at each point $M \in S$; given two unit vectors $\vec{e}_1$ and $\vec{e}_2$, tangential to S at M and orthogonal, an orientation of S in the neighbourhood of M is defined. Given such a system $(\vec{e}_1, \vec{e}_2)$, there exists a unique unit vector $\vec{e}_3$, orthogonal to $\vec{e}_1$ and $\vec{e}_2$, such that $\det(\vec{e}_1, \vec{e}_2, \vec{e}_3) = +1$; conversely, the choice of a unit vector $\vec{e}_3$, normal to the surface S at the point $M \in S$, defines an orientation of S in the neighbourhood of M: we choose $\vec{e}_1$ and $\vec{e}_2$ such that $\det(\vec{e}_1, \vec{e}_2, \vec{e}_3) = +1$. Thus, to orientate S, it suffices to *choose at each point* $M \in S$ *one of the two unit vectors normal to* S, *this choice varying continuously with* M. In future we shall suppose that S has been *oriented* (which implies that S is "orientable").

Thus, let C be an oriented curve traced on an oriented surface S in $\mathbf{R}^3$. We shall associate with each point $M \in C$ a direct orthonormal frame with origin M, called the

Darboux frame (of C *vis-à-vis* S). For $\vec{e}_1$ we take the unit vector tangential to C at the point M, as in the case of the Frenet frame. Next $\vec{e}_2$ is taken tangential to S at the point M and orthogonal to $\vec{e}_1$, and such that det $(\vec{e}_1, \vec{e}_2, \vec{e}_3) = +1$, where $\vec{e}_3$ is the unit normal to S which defines the orientation of S. Thus the curve C traced on S defines a parameter-dependent family of frames, i.e. a curve in the variety *SO*. The equations of motion of the frame are

$$(1.5.1) \qquad \begin{cases} dM = \omega_1 \vec{e}_1, \\ d\vec{e}_1 = \omega_{12}\vec{e}_2 + \omega_{13}\vec{e}_3, \qquad de_2 = -\omega_{12}\vec{e}_1 + \omega_{23}\vec{e}_3 \\ d\vec{e}_3 = -\omega_{13}\vec{e}_1 - \omega_{23}\vec{e}_2. \end{cases}$$

If we take the arc length s of the oriented curve C as parameter, we have $\omega_1 = ds$, $\omega_{12} = a\,ds$, $\omega_{13} = b\,ds$, $\omega_{23} = c\,ds$, where a, b, c are functions on C. They are called, respectively:

the *geodesic curvature*
the *normal curvature*
the *geodesic torsion*

of the curve C (they will be explained below). We can therefore write

$$(1.5.2) \qquad \begin{cases} \dfrac{\omega_{12}}{ds} = \text{geodesic curvature} \\[2ex] \dfrac{\omega_{13}}{ds} = \text{normal curvature} \\[2ex] \dfrac{\omega_{23}}{ds} = \text{geodesic torsion.} \end{cases}$$

PROPOSITION 1.5.1. *If the orientation of* C *is changed, the normal curvature and the geodesic torsion are unchanged, and the geodesic curvature is multiplied by* -1.

Indeed, $\omega_1 = ds$ is multiplied by -1, as well as $\vec{e}_1$ and $\vec{e}_2$; $\vec{e}_3$ is unchanged; the relations (1.5.3) then show that ω_{12} is unchanged, whereas ω_{13} and ω_{23} are multiplied by -1. Hence the result.

PROPOSITION 1.5.2. Two curves traced on S and tangential to each other at a point $M \in S$, have the *same normal curvature* and the *same geodesic torsion* at M. (In other words, the normal curvature and the geodesic curvature are associated with each tangential direction at the point $M \in S$.)

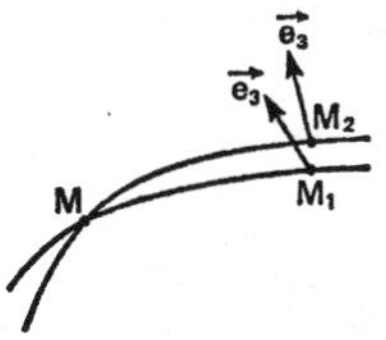

PROOF. For two tangential curves C_1 and C_2, $\vec{e}_1$ and $\vec{e}_2$ have the same values at the point M; this is also true for $d\vec{e}_3/ds$. (For the vectors $\vec{e}_3$ normal to S at the points $M_1 \in C_1$ and

$M_2 \in C_2$ corresponding to the same value of the arc length s, measured from M, are equal to the first order of infinitesimals: their difference is negligible *vis-à-vis* s. In fact the distance between M_1 and M_2 is an infinitesimal of the second order, and the normal vector $\vec{e}_3$ at a point of S is a differentiable function of this point.) Then the last of relations (1.5.1) gives

$$\frac{d\vec{e}_3}{ds} = -\frac{\omega_{13}}{ds}\vec{e}_1 - \frac{\omega_{23}}{ds}\vec{e}_2,$$

which shows that ω_{13}/ds and ω_{23}/ds are the same for the two curves at the point M.

Q.E.D.

Remark. It is *not true* that the geodesic curvature is the same for two curves traced on S and tangential at M. For example, if S is a *plane*, we see that the geodesic curvature of a plane curve is none other than the (ordinary) curvature; it is well known that two plane curves, tangential at a common point M do not necessarily have the same curvature at M.

1.6 *Calculation of the geodesic curvature, normal curvature and geodesic torsion*

As before let C be an oriented curve traced on an oriented surface S in $\mathbf{R}^3$. Let $(\vec{e}_1, \vec{e}_2, \vec{e}_3)$ be the Darboux frame at the point M of C. Denote by $(\vec{e}_1, \vec{n}, \vec{b})$ the Frenet frame of C at the point M: the unit vector $\vec{n}$ is called the "principal normal" of C and $\vec{b}$ is the "binormal". Let θ denote the angle $(\vec{n}, \vec{e}_3)$ between the vector $\vec{e}_3$ normal to the surface and the vector $\vec{n}$.

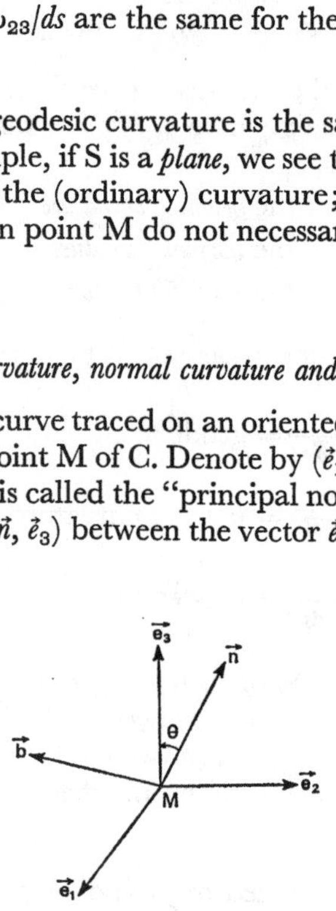

(1.6.1)
$$\begin{cases} \vec{n} = \vec{e}_2 \sin\theta + \vec{e}_3 \cos\theta \\ \vec{b} = -\vec{e}_2 \cos\theta + \vec{e}_3 \sin\theta, \end{cases}$$

or, inversely,

(1.6.2)
$$\begin{cases} \vec{e}_2 = \vec{n} \sin\theta - \vec{b} \cos\theta \\ \vec{e}_3 = \vec{n} \cos\theta + \vec{b} \sin\theta. \end{cases}$$

By the equations of motion of the Frenet frame, we have

$$\frac{d\vec{e}_1}{ds} = \frac{1}{\rho}\vec{n} = \frac{\sin\theta}{\rho}\vec{e}_2 + \frac{\cos\theta}{\rho}\vec{e}_3,$$

hence, by comparing with (1.5.1):

(1.6.3)
$$\begin{cases} \boxed{\text{geodesic curvature} = \dfrac{\sin\theta}{\rho} \\ \text{normal curvature} = \dfrac{\cos\theta}{\rho}} \end{cases}$$

(ρ denotes the radius of curvature of the curve C).

It remains to calculate the geodesic torsion ω_{23}/ds. By (1.5.1) we have

$$\frac{\omega_{23}}{ds} = \vec{e}_3 \cdot \frac{d\vec{e}_2}{ds} = (\vec{n}\cos\theta + \vec{b}\sin\theta) \cdot \frac{d}{ds}(\vec{n}\sin\theta - \vec{b}\cos\theta).$$

By taking account of the equations of motion of the Frenet frame:

$$\frac{d\vec{n}}{ds} = -\frac{1}{\rho}\vec{e}_1 + \frac{1}{\tau}\vec{b}, \qquad \frac{d\vec{b}}{ds} = -\frac{1}{\tau}\vec{n},$$

we obtain

$$\vec{e}_3 \frac{d\vec{e}_2}{ds} = \frac{1}{\tau} + \frac{d\theta}{ds}.$$

Thus

(1.6.4)
$$\boxed{\text{geodesic torsion} = \frac{1}{\tau} + \frac{d\theta}{ds}}.$$

Exercise. Meusnier's theorem. This states that the "normal curvature" $(\cos\theta)/\rho$ only depends on the tangent of the curve C at the point $M \in C$. Thus: (1) if two curves lie on S and pass through M, where they have the same tangent and the same osculating plane, then they have the same radius of curvature (since $\cos\theta$ is the same for both curves at M); (2) if a plane P pivots about a line D tangential to S at a point $M \in S$, then the centre of curvature (at the point M) of the section of S by P describes a circle passing through M.

Remarks. (i) Let us cut S with a plane orthogonal to the tangent plane to S at the point $M \in S$; let C be the curve of intersection of this plane with S. On this curve, $\theta = 0$, therefore the curvature $1/\rho$ of C is equal to the "normal curvature" in the direction of the tangent to C at the point M. Thus the "normal curvature" is the *curvature of the normal section* (section of S made by the normal plane which passes through the tangent under consideration).

(ii) We know that through a point $M \in S$ there passes one and only one geodesic of S which is tangential to a given straight line (tangential to S). Along a geodesic we have either $\theta = 0$ or π, so that $d\theta/ds = 0$; the geodesic torsion of such a curve is therefore equal to $1/\tau$. Thus: *the geodesic torsion in a direction tangential to S is equal to the torsion of the geodesic tangential to this direction.*

(iii) To say that a curve is a geodesic, is the same as saying that $\sin\theta = 0$; therefore, *the geodesics are the curves on S with vanishing geodesic curvature.*

(iv) If S is a plane, and C is a curve lying in the plane, then $\theta = \pi/2$ (for a suitable choice of the vector $\vec{n}$), therefore, $(\sin\theta)/\rho = 1/\rho$: the geodesic curvature is none other than the curvature of C.

2. 3-Parameter family of frames associated with a surface in $\mathbf{R}^3$

2.1 *The variety of frames of an oriented surface*

Let $S \subset \mathbf{R}^3$ be a surface of class k (k sufficiently large) supposed *oriented.* Let R(S) denote the set of all direct orthonormal frames whose origin M belongs to S, and whose vector $\vec{e}_3$ is normal to S (which defines the orientation of S). If we associate each frame with its origin, we define a continuous mapping

$$p: \mathrm{R(S)} \to \mathrm{S},$$

evidently surjective. The *fibre* of a point $\mathrm{M} \in \mathrm{S}$, i.e. the inverse image $p^{-1}(\mathrm{M})$ consists of all frames with origin M; such a frame is defined by the choice of a unit vector $\vec{e}_1$ tangential to S at the point M (for, as soon as $\vec{e}_1$ is known, $\vec{e}_2$ is determined since $(\vec{e}_1, \vec{e}_2, \vec{e}_3)$ must be direct). All frames with origin M can be obtained one from another by rotation about the vector $\vec{e}_3$. More precisely, let us consider the group of rotations S0(2) of the plane about the origin (group of real linear transformations of two variables, orthogonal and with determinant $+1$); an element of this group is defined by an angle φ (a real number modulo 2π). The group S0(2) *operates in the space* **R**(S) *of frames*: given a frame $(\mathrm{M}, \vec{e}_1, \vec{e}_2, \vec{e}_3)$ we know how to make a rotation through an angle φ about $\vec{e}_3$; we obtain a frame $(\mathrm{M}', \vec{e}_1{}', \vec{e}_2{}', \vec{e}_3{}')$ defined by

$$(2.1.1) \qquad \begin{cases} \mathrm{M}' = \mathrm{M}, & \vec{e}_1{}' = \vec{e}_1 \cos\varphi + \vec{e}_2 \sin\varphi \\ \vec{e}_2{}' = -\vec{e}_1 \sin\varphi + \vec{e}_2 \cos\varphi, & \vec{e}_3{}' = \vec{e}_3. \end{cases}$$

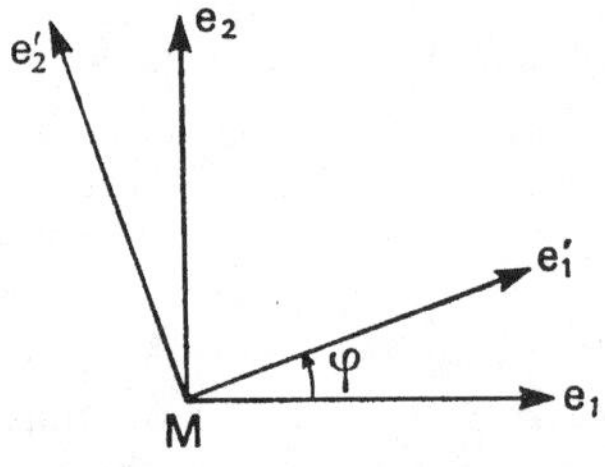

We see that the *fibres* of the mapping p: R(S) → S are the *orbits* of the group S0(2) operating in R(S), and that S0(2) operates in a *simply transitive* manner in each fibre (given two frames with the same origin, there exists a unique rotation which transforms the first into the second).

The mapping p: R(S) → S furnishes an example of what is known as a "fibre space"; we shall not consider the theory of such spaces here, nor even their definition.

Let us now show how we can parametrize the set of frames of R(S) whose origin M is sufficiently close to a given point $\mathrm{M}_0 \in \mathrm{S}$. Let us suppose that a vertical line through the point $\mathrm{M}_0 \in \mathbf{R}^3$ is not tangential to S at M_0, so that, in the neighbourhood of M_0, S may be represented by an equation

$$z = f(x, y), \qquad f \text{ of class } \mathrm{C}^k$$

(where x, y, z are coordinates in $\mathbf{R}^3$). Then, at each point M in the neighbourhood of M_0, there exists a unique unit vector tangential to S whose projection on the plane $z = 0$ is parallel to the vector $(1, 0, 0)$; this vector $\vec{e}_1(M)$ is a function of M of class C^{k-1}. Thus we associate with each point $M \in S$, in the neighbourhood of M_0, an orthonormal frame having M for origin; in other words, we have defined a continuous mapping

$$\sigma: S \to R(S)$$

such that $p \circ \sigma =$ identity of S; σ is known as a *section* of the fibre space of frames. Then each frame $r \in R(S)$, with origin M in the neighbourhood of M_0, is defined by: (1) The point $M \in S$; (2) the angle φ through which the frame

$$(M, \vec{e}_1(M), \vec{e}_2(M), \vec{e}_3(M))$$

must be rotated to make it coincide with r. In other words, the space of frames with origin M lying in a sufficiently small neighbourhood V of M_0 can be represented by a point in the product space $V \times SO(2)$. In particular, these frames depend on three real parameters. We can show that they constitute a *sub-variety* of dimension 3, of class C^{k-1}, of the space of *all* frames (not necessarily orthonormal) of $\mathbf{R}^3$ (which, themselves, form an open subset of $\mathbf{R}^{12}$; cf. § 1). (This is left as an exercise for the reader.) Then the mapping $p: R(S) \to S$ is of class C^{k-1}; the section $\sigma: S \to R(S)$ defined above is of class C^{k-1}.

If α is a differential form on the surface S, $p^*(\alpha)$ is a differential form on the variety R(S). Let us recall that a differential form on R(S) is a function $\omega(r, \tau)$, [where $r \in R(S)$ and τ is tangential to R(S) at r], which is linear in τ for each r; in order that this differential form be of the type $p^*(\alpha)$, it is necessary and sufficient that: (1) for each r, $\omega(r, \tau)$ depends only on the projection $\xi = p(\tau)$ of the vector, τ, which is a vector tangential to S at the point $M = p(r)$; (2) $\omega(r, \xi)$ depends only on the origin $M = p(r)$ of the frame r. We then say, for brevity, that ω *is a differential form of the surface* S.

These notions will be illustrated later by some examples.

2.2 *The equations of motion of a frame associated with an oriented surface*

In the case of the 3-parameter family R(S) of orthonormal frames, the equations of motion of a frame introduce (cf. (1.1.2), (1.1.3), (1.3.2) and (1.3.3)) differential forms $\omega_1, \omega_2, \omega_3, \omega_{12}, \omega_{13}, \omega_{23}$ on the variety R(S). But $\omega_3 = 0$, since, for each $r = (M, \vec{e}_1, \vec{e}_2, \vec{e}_3)$, the differential form dM has values in the tangent plane of S, generated by the vectors $\vec{e}_1$ and $\vec{e}_2$ of the frame r. Thus the equations of motion are:

(2.2.1) $$dM = \omega_1\vec{e}_1 + \omega_2\vec{e}_2$$

(2.2.2) $$\begin{cases} d\vec{e}_1 = \omega_{12}\vec{e}_2 + \omega_{13}\vec{e}_3 \\ d\vec{e}_2 = -\omega_{12}\vec{e}_1 + \omega_{23}\vec{e}_3 \\ -d\vec{e}_3 = \omega_{13}\vec{e}_1 + \omega_{23}\vec{e}_2. \end{cases}$$

On the other hand, the conditions of integrability (cf. (1.2.1) and (1.2.2)) give in particular (taking account of the fact that $\omega_3 = 0$):

(2.2.3) $$d\omega_1 = -\omega_2 \wedge \omega_{12}, \qquad d\omega_2 = \omega_1 \wedge \omega_{12},$$

(2.2.4) $$\omega_1 \wedge \omega_{13} + \omega_2 \wedge \omega_{23} = 0,$$

(2.2.5) $$d\omega_{12} = -\omega_{13} \wedge \omega_{23}.$$

Interpretation of ω_1 *and* ω_2. Let us recall that the differential dM is the differential 1-form with values in $\mathbf{R}^3$ which, to each vector $\vec{\xi}$, associates the same vector $\vec{\xi}$. Thus, if r is a frame $\in R(S)$, and if $\vec{e}_1(r)$ and $\vec{e}_2(r)$ are the first two vectors of this frame, the differential 1-forms $\omega_1(r, \xi)$ and $\omega_2(r, \xi)$ are linear forms which associate to each vector $\vec{\xi}$ tangential to S at the origin $M(r)$ of the frame its coordinates relative to the base $\vec{e}_1(r)$ and $\vec{e}_2(r)$. Considered as differential forms on the variety of frames R(S), ω_1 and ω_2 are therefore functions of $r \in R(S)$ which, for each r, depend only on the vector ξ tangential to S at the point $M(r) \in S$:

(2.2.6) $$\vec{\xi} = \omega_1(r, \xi)\vec{e}_1(r) + \omega_2(r, \xi)\vec{e}_2(r).$$

The square of the vector $\vec{\xi}$ tangential to S at the point M is therefore equal to

$$|\vec{\xi}|^2 = (\omega_1(r, \xi))^2 + (\omega_2(r, \xi))^2.$$

We have been led to introduce the notation $(\omega_1)^2$ for the *square* of the differential form ω_1; this is not a differential 2-form in the sense considered here (i.e., a function of r with values in the space of *bilinear alternating forms* on the tangent space to S at the point $M(r)$), but it is a function of r with values in the space of *quadratic forms* on this tangent space: $(\omega_1(r, \xi))^2$ is the square of the linear form $\omega_1(r, \xi)$.

DEFINITION. The quadratic differential form $(\omega_1)^2 + (\omega_2)^2$ is called *the first fundamental quadratic form of the surface* S. With each point $M \in S$, and with each tangent vector $\vec{\xi}$ of S at the point M, it associates the square of the length of $\vec{\xi}$. It does not depend on the frame with origin M, but only on the origin.

This quadratic form is often called the ds^2 of the surface, the reason being that, if we consider an oriented surface C traced on S, the form induced on C is the square of the differential 1-form ds of the curve C. When the coordinates x, y, z are expressed as function of two parameters u and v, we have

$$ds^2 = E(u, v)\, du^2 + 2F(u, v)\, du\, dv + G(u, v)\, dv^2$$

with,

(2.2.7) $$\begin{cases} E = \left(\dfrac{\partial x}{\partial u}\right)^2 + \left(\dfrac{\partial y}{\partial u}\right)^2 + \left(\dfrac{\partial z}{\partial u}\right)^2 \\ F = \dfrac{\partial x}{\partial u}\dfrac{\partial x}{\partial v} + \dfrac{\partial y}{\partial u}\cdot\dfrac{\partial y}{\partial v} + \dfrac{\partial z}{\partial u}\cdot\dfrac{\partial z}{\partial v} \\ G = \left(\dfrac{\partial x}{\partial v}\right)^2 + \left(\dfrac{\partial y}{\partial v}\right)^2 + \left(\dfrac{\partial z}{\partial v}\right)^2. \end{cases}$$

2.3 *The element of area of the surface*

S being oriented, the element of area is a differential 2-form (cf. Chap. 1, § 4.12), namely that which, with a pair of vectors ξ_1 and ξ_2 tangential to S at the point M, associates

$$\det(\vec{\xi}_1, \vec{\xi}_2, \vec{e}_3).$$

By (2.2.6) this is equal to

$$\omega_1(\xi_1)\omega_2(\xi_2) - \omega_1(\xi_2)\omega_2(\xi_1).$$

Now, for the pair (ξ_1, ξ_2), this is the value of the differential form

$$\omega_1 \wedge \omega_2,$$

which is the exterior product of ω_1 and ω_2. Thus:

PROPOSITION 2.3.1. *The form $\omega_1 \wedge \omega_2$ is equal to the element of area of the surface.*

We have already seen in Chap. 1 (equation 4.12.6) that this differential form is equal to $\sqrt{EG - F^2}\, du \wedge dv$, when E, F, G are the coefficients of ds^2 of the surface S parametrized by u and v in accordance with the orientation of S.

Similarly to the quadratic form $\omega_1^2 + \omega_2^2$, the form $\omega_1 \wedge \omega_2$ is a differential form of the surface S (cf. end of § 2.1).

2.4 *The second fundamental quadratic form of the surface* S

The differential forms dM and $-d\vec{e}_3$ (with values in $\mathbf{R}^3$) are, for each $M \in S$, linear functions of the vector $\vec{\xi}$ tangential to S at the point M. Their *scalar product*

$$(d\vec{M})\cdot(-d\vec{e}_3)$$

is a quadratic form of the vector ξ of the tangent space to S. By definition, it is the *second fundamental quadratic form of the surface* S. It is easy to calculate: by (2.2.1) and (2.2.2) we have

$$(2.4.1) \qquad \boxed{(d\vec{M})\cdot(-d\vec{e}_3) = \omega_1\omega_{13} + \omega_2\omega_{23}}.$$

(On the right-hand side, $\omega_1\omega_{13}$ *must not be confused with the exterior product* $\omega_1 \wedge \omega_{13}$; for each $r \in R(S)$, and for each vector τ tangential to R(S) at the point r, $\omega_1\omega_{13}$ is equal to the product

$$\omega_1(r, \tau)\omega_{13}(r, \tau),$$

whereas, for each r, and each pair (τ_1, τ_2), $\omega_1 \wedge \omega_{13}$ is equal to

$$\omega_1(r, \tau_1)\omega_{13}(r, \tau_2) - \omega_1(r, \tau_2)\omega_{13}(r, \tau_1).)$$

Interpretation of the second fundamental quadratic form in terms of the normal curvature. Let C be an oriented curve traced on S, and parametrized by its arc length s. The equations of motion of the Darboux frame (cf. (1.5.1)) show that the scalar product

$$\frac{d\vec{M}}{ds}\cdot\left(-\frac{d\vec{e}_3}{ds}\right)$$

is equal to the *normal curvature* of C at the point $M \in C$; this signifies that the value of the second fundamental form on the vector $\vec{\xi}$ tangential to C is equal to the product of $|\vec{\xi}|^2$ and the normal curvature of C. Hence:

PROPOSITION 2.4.1. *At a point* $M \in S$, *the value of the second fundamental quadratic form on a* *vector $\vec{\xi} \neq 0$, tangential to S at the point M, is equal to the product of $|\vec{\xi}|^2$ and the normal curvature in the direction of the vector $\vec{\xi}$.*

2.5 *Calculation of the normal curvature and geodesic torsion in a given direction*

After exterior multiplication by ω_1, relation (2.2.4) gives:

$$\omega_1 \wedge \omega_2 \wedge \omega_{23} = 0,$$

which shows that, for each $r \in R(S)$, the three linear forms $\omega_1, \omega_2, \omega_{23}$ (on the tangent space of R(S)) are linearly dependent. Now ω_1 and ω_2 are linearly independent by (2.2.6). Thus ω_{23} is equal to a linear combination of ω_1 and ω_2:

$$\omega_{23} = b\omega_1 + c\omega_2,$$

where b and c are functions of the frame r. In particular, this shows that for each frame r, ω_{23} is in fact a linear form on the tangent space of S at the point $M(r)$.

Similarly, we have

$$\omega_{13} = a\omega_1 + b'\omega_2,$$

where a and b' are functions of r. But, substituting in (2.2.4), we have

$$b'\omega_1 \wedge \omega_2 + b\omega_2 \wedge \omega_1 = 0,$$

hence, since the form $\omega_1 \wedge \omega_2$ is not zero,

$$\boxed{b = b'}.$$

Thus, we have

$$\boxed{\omega_{13} = a\omega_1 + b\omega_2, \qquad \omega_{23} = b\omega_1 + c\omega_2}; \tag{2.5.1}$$

and we have introduced *three functions a, b, c on the space R(S) of frames*. We shall now interpret these functions.

First of all, the relation (2.4.1) tells us that the *second fundamental quadratic form is equal to*

$$\boxed{\omega_1\omega_{13} + \omega_2\omega_{23} = a(\omega_1)^2 + 2b\omega_1\omega_2 + c(\omega_2)^2}. \tag{2.5.2}$$

If a vector $\vec{\xi}$ tangential to S at the point $M(r)$ makes an angle φ with the vector $\vec{e}_1(r)$ of the frame r, then evidently we have

$$\omega_1(r, \vec{\xi}) = |\vec{\xi}| \cos \varphi, \qquad \omega_2(r, \vec{\xi}) = |\vec{\xi}| \sin \varphi;$$

thus the value of the second fundamental quadratic form, for the vector $\vec{\xi}$, is

$$|\vec{\xi}|^2(a \cos^2 \varphi + 2b \sin \varphi \cos \varphi + c \sin^2 \varphi).$$

By Prop. 2.4.1., we obtain:

PROPOSITION 2.5.1. *The normal curvature in the direction which makes an angle φ with the vector $\vec{e}_1$ of the frame r is equal to*

$$\boxed{a(r) \cos^2 \varphi + 2b(r) \sin \varphi \cos \varphi + c(r) \sin^2 \varphi}. \tag{2.5.3}$$

COROLLARY. $a(r)$ is the normal curvature in the direction of the vector $\vec{e}_1$ of the frame r, and $c(r)$ is the normal curvature in the direction of the vector $\vec{e}_2$ of the frame r.

We have thus interpreted the functions a and c. It remains to interpret b. To do this we shall calculate the *geodesic torsion* in the direction which makes an angle φ with the vector $\vec{e}_1$ of the frame r. Consider a curve C traced on S and parametrized by its arc length s; let $M \in C$, let $r = (M, \vec{e}_1, \vec{e}_2)$ be a frame with origin M, and let φ be the angle between the unit vector $\vec{e}'_1$ tangential to C and $\vec{e}_1$; let $\vec{e}'_2$ be the unit tangent vector making an angle $+\pi/2$ with $\vec{e}'_1$. The equations of motion of the Darboux frame $(M, \vec{e}'_1, \vec{e}'_2, \vec{e}'_3)$ show that the geodesic torsion of C is equal to the coefficient of $\vec{e}'_2$ in the expression

$$-\frac{d\vec{e}_3}{ds} = \alpha\vec{e}'_1 + \beta\vec{e}'_2.$$

In the third of relations (2.2.2), replace $\vec{e}_1$ by $\vec{e}'_1 \cos\varphi - \vec{e}'_2 \sin\varphi$, and $\vec{e}_2$ by $\vec{e}'_1 \sin\varphi + \vec{e}'_2 \cos\varphi$ (cf. (2.1.1)); then

$$\beta = -\frac{\omega_{13}}{ds}\sin\varphi + \frac{\omega_{23}}{ds}\cos\varphi;$$

substituting from (2.5.1) for ω_{13} and ω_{23}, and replacing ω_1 by $ds \cos\varphi$ and ω_2 by $ds \sin\varphi$, we find:

$$\begin{aligned}\beta &= -(a\cos\varphi + b\sin\varphi)\sin\varphi + (b\cos\varphi + c\sin\varphi)\cos\varphi\\ &= b(\cos^2\varphi - \sin^2\varphi) + (c - a)\sin\varphi\cos\varphi.\end{aligned}$$

Hence:

PROPOSITION 2.5.2. *The geodesic torsion in a direction making an angle φ with the vector $\vec{e}_1$ of the frame r is equal to*

$$\boxed{b\cos 2\varphi + \frac{c - a}{2}\sin 2\varphi}. \tag{2.5.4}$$

COROLLARY. $b(r)$ is the geodesic torsion in the direction of the vector $\vec{e}_1$ of the frame r; the geodesic torsion in the direction of the vector $\vec{e}_2$ is equal to $-b(r)$ (set $\varphi = \pi/2$).

More generally, changing φ to $\varphi + \pi/2$ in (2.5.4) we obtain:

PROPOSITION 2.5.3. *At a point* $M \in S$, *the values of the geodesic torsion in two orthogonal directions are equal except for a sign change.*

Referring to equation (2.5.3), changing φ to $\varphi + \pi/2$ shows that *the sum of the normal curvatures in the directions φ and $\varphi + \pi/2$ is independent of φ.* In other words:

PROPOSITION 2.5.4. The sum $a(r) + c(r)$ of the normal curvatures in the directions of the vectors $\vec{e}_1(r)$ and $\vec{e}_2(r)$ of the frame r *depends only on the origin* M *of the frame* r.

DEFINITION. $a(r) + c(r)$ is called the mean curvature of the surface S at the point M.

2.6 *Principal directions; lines of curvature*

In order that the geodesic torsion be null in every direction tangential to S at the point M, it is necessary and sufficient that, for a particular frame r with origin M, the two quantities $b(r)$ and $c(r) - a(r)$ be zero: this follows from the expression (2.5.4) for the geodesic torsion. If this is the case, M is said to be an *umbilical point* of the surface S. And then the normal curvature is the same in every tangential direction at the point M; this follows from (2.5.3). *Exercise:* conversely, if the normal curvature is the same in all tangential directions at M, M is an umbilical point.

PROPOSITION 2.6.1. If M is not an umbilical point of the surface S, there exist precisely *two tangential directions for which the geodesic torsion is zero;* they are at *right angles.*

This is evident from (2.5.4): in order that the geodesic torsion be zero in a direction φ, it is necessary and sufficient that

$$\tan 2\varphi = \frac{2b}{a - c},$$

which gives a well defined value (finite or infinite) of $\tan 2\varphi$, since b and $a - c$ do not vanish simultaneously. Hence the result.

DEFINITION. The two (orthogonal) directions in which the geodesic torsion vanishes are called the *principal directions* (at the point $M \in S$ considered). At an umbilical point, we agree that all directions are principal.

In the neighbourhood of a point $M_0 \in S$ which is not an umbilical point, the principal directions at every point M constitute two fields of tangent vectors.

DEFINITION. A *line of curvature* of a surface S is a curve C traced on S such that, at every point $M \in C$, the tangent to C is in a principal direction.

If we suppose that S is of class C^k (k sufficiently large), the fields of principal directions are of class C^1; we can therefore apply the theory of differential equations to deduce that, *through a given point* M_0 (other than an umbilical point) *pass two mutually orthogonal lines of curvature.*

It is evident that at every point of a line of curvature the geodesic torsion is zero. In the notation of § 1.6 (equation (1.6.4)), this is expressed by

$$\frac{1}{\tau} + \frac{d\theta}{ds} = 0, \tag{2.6.1}$$

$1/\tau$ denoting the torsion of the line of curvature, and θ the angle between the normal $\vec{e}_3$ to the surface and the principal normal of the line of curvature. *Exercise:* condition (2.6.1) expresses that, when M describes a line of curvature C, the normal to S at every point of C, generates a "developable surface" or again, that this normal remains tangential to a fixed curve.

We often denote by $1/R_1$ and $1/R_2$ the values of the normal curvature (at a point $M \in S$) in the two principal tangential directions at M. Naturally we cannot distinguish between R_1 and R_2. The numbers $1/R_1$ and $1/R_2$ are called the *principal curvatures* at the point M. For an umbilical point, we agree that $1/R_1$ and $1/R_2$ are equal, and equal to the normal curvature in an arbitrary direction. Prop. 2.5.4 says that *the sum of the normal curvatures in two orthogonal* directions is equal to $1/R_1 + 1/R_2$ (*mean curvature* at the point M).

Consider at the point M a frame whose vectors $\vec{e}_1$ and $\vec{e}_2$ are tangential to the principal directions. The expression (2.5.3) shows that the normal curvature in a direction which makes an angle φ with $\vec{e}_1$ is equal to

$$\boxed{\frac{1}{R_1}\cos^2\varphi + \frac{1}{R_2}\sin^2\varphi}$$

(where $1/R_1$ is the normal curvature in the direction $\vec{e}_1$, and $1/R_2$ is the normal curvature in the direction of $\vec{e}_2$). If r is a frame making an angle φ with the preceding one, we therefore have,

$$a(r) = \frac{1}{R_1}\cos^2\varphi + \frac{1}{R_2}\sin^2\varphi$$

$$c(r) = \frac{1}{R_1}\sin^2\varphi + \frac{1}{R_2}\cos^2\varphi$$

$$b(r) = \left(\frac{1}{R_2} - \frac{1}{R_1}\right)\sin\varphi\cos\varphi \qquad \text{(by (2.5.4))}.$$

It follows that

$$a(r)c(r) - b(r)^2 = \frac{1}{R_1R_2}.$$

Hence:

PROPOSITION 2.6.2. The quantity $ac - b^2$ associated with each frame r depends only on the origin M of the frame, and is equal to the product of the principal curvatures $1/R_1$ and $1/R_2$. It is called the *total curvature* of the surface S at the point M.

2.7 *The differential form of geodesic curvature*

We consider the differential form ω_{12}: let us recall that if C is a curve traced on S, and $\Gamma \in R(S)$ is the curve defined by the Darboux frames associated with the points of C, the differential form induced on Γ by ω_{12} is equal to ds/ρ_g, where ds denotes the element of arc of C, and $1/\rho_g$ the geodesic curvature. The curve Γ is called the "canonical lifting" of C in R(S).

THEOREM 2.7.1. *The differential form of geodesic curvature ω_{12} is the unique differential form of degree one (on* R(S)*) which satisfies equations* (2.2.3):

$$d\omega_1 = -\omega_2 \wedge \omega_{12}, \qquad d\omega_2 = \omega_1 \wedge \omega_{12}.$$

PROOF. If there exists a second form ω'_{12} satisfying the same equations, the form $\alpha = \omega_{12} - \omega'_{12}$ satisfies

$$\omega_2 \wedge \alpha = 0, \qquad \omega_1 \wedge \alpha = 0.$$

This means that α is proportional to ω_2, and also proportional to ω_1. Now ω_1 and ω_2 are not proportional. Thus $\alpha = 0$. Q.E.D.

We shall show that the *geodesic curvature* of a curve C traced on S is an invariant of the ds^2 of the surface. More precisely:

PROPOSITION 2.7.2. Let f: S $\to$ S′ be a diffeomorphism of a surface S onto a surface S′.

Let us suppose that F preserves lengths (i.e., the ds^2 of S is transformed into the ds^2 of S'). Then, if C is an oriented curve traced on S, and if $C' = f(C)$, the geodesic curvature of C at a point $M \in C$ is equal to the geodesic curvature of C' at the point $f(M)$.

PROOF. With each orthonormal frame $\in R(S)$, f associates an orthonormal frame $\in R(S')$, since the derivative f' transforms every tangent vector to S into a tangent vector to S' *with the same length* (by hypothesis). It follows that the change of variable f transforms the forms ω_1' and ω_2' (of R(S')) into the forms ω_1 and ω_2 (of R(S)). This change of variables therefore transforms $d\omega_1'$ into $d\omega_1$, and $d\omega_2'$ into $d\omega_2$. That ω_{12}' is transformed into ω_{12} follows from the uniqueness Theorem 2.7.1.

Q.E.D.

We shall now use relation (2.2.5) with ω_{13} and ω_{23} replaced by their values from (2.5.1). We obtain:

$$\begin{aligned} d\omega_{12} &= -(a\omega_1 + b\omega_2) \wedge (b\omega_1 + c\omega_2) \\ &= -(ac - b^2)\omega_1 \wedge \omega_2. \end{aligned}$$

We have seen that $\omega_1 \wedge \omega_2$ is *the element of area*; in the following we shall denote it by $d\sigma$ (a differential form of degree 2). On the other hand, $ac - b^2$ is none other than the *total curvature* (Prop. 2.6.2). Thus:

THEOREM 2.7.3. *The exterior differential $d\omega_{12}$ of the differential form of geodesic curvature is equal to the differential 2-form $-d\sigma/R_1R_2$, minus the product of the area element and the total curvature.*

2.8 *Use of a field of frames*

Let us suppose that on S there exists a "field of frames", i.e., a section $r: S \to R(S)$ (cf. § 2.1) which is sufficiently differentiable. The frame $r(M)$ associated with a point $M \in S$ is defined by a vector $\vec{e}_1(M)$, of unit length and tangential to S at the point M (and the vector $\vec{e}_2(M)$ of the frame $r(M)$ makes an angle $+\pi/2$ with $\vec{e}_1(M)$). By virtue of this field of frames, R(S) is identified with the product $S \times SO(2)$, as we have seen in § 2.1: every frame r is determined by its origin $M \in S$ and by the angle φ between its first vector $\vec{e}_1$ and the vector $\vec{e}_1(M)$ of the field.

Observe that such a field of frames exists each time that we have a parametrization of S, i.e. a diffeomorphism Φ of an open $U \subset \mathbf{R}^2$ (with coordinates u, v) on S. Indeed, the derived mapping $\Phi'(u, v)$ transforms the vector (1, 0) of the (u, v)-plane into a vector $\vec{\xi}$ tangential to S at the point $M = \Phi(u, v)$; $\vec{\xi}$ is $\neq 0$, and defines a unique unit vector $\vec{e}_1(M)$, proportional to $\vec{\xi}$ (with the coefficient of proportionality > 0).

By means of a field of frames, S is identified with a sub-variety of $R(S) = S \times SO(2)$, viz. the sub-variety of the frames of the field. Let $\bar{\omega}_1$, $\bar{\omega}_2$ and $\bar{\omega}_{12}$ be the differential forms induced on S by ω_1, ω_2 and ω_{12}. By (2.2.3) we have

$$(2.8.1) \qquad d\bar{\omega}_1 = -\bar{\omega}_2 \wedge \bar{\omega}_{12}, \qquad d\bar{\omega}_2 = \bar{\omega}_1 \wedge \bar{\omega}_{12}.$$

We shall see later (§ 2.11) that these relations enable us to determine $\bar{\omega}_{12}$ if $\bar{\omega}_1$ and $\bar{\omega}_2$ are known.

Lemma 2.8.1.

$$\boxed{\omega_{12} = \bar{\omega}_{12} + d\varphi}.$$

PROOF. This follows from the formulae:

$$\begin{cases} \vec{e}_1 = \vec{e}_1(\mathrm{M}) \cos \varphi + \vec{e}_2(\mathrm{M}) \sin \varphi \\ \vec{e}_2 = -\vec{e}_1(\mathrm{M}) \sin \varphi + \vec{e}_2(\mathrm{M}) \cos \varphi \\ \omega_{12} = \vec{e}_2 \cdot d\vec{e}_1 \\ \bar{\omega}_{12} = \vec{e}_2(\mathrm{M}) \cdot d\vec{e}_1(\mathrm{M}) = -\vec{e}_1(\mathrm{M}) \cdot d\vec{e}_2(\mathrm{M}) \end{cases}$$

and a simple calculation.

As a consequence of the lemma we have,

$$d\bar{\omega}_{12} = d\omega_{12} = -\frac{d\sigma}{\mathrm{R}_1\mathrm{R}_2}. \tag{2.8.2}$$

2.9 *Parallel transport along a curve*

We again take the hypotheses of § 2.8 (existence of a field of frames). Let Γ be an oriented curve of class C^1 in R(S); it is defined (1) by the prescription of an oriented curve C traced on S, viz. the position of the origin of the frame; (2) by giving, for each $\mathrm{M} \in \mathrm{C}$, the angle φ between the frame of Γ and the frame $(\vec{e}_1(\mathrm{M}), \vec{e}_2(\mathrm{M}))$ of the field at the point M.

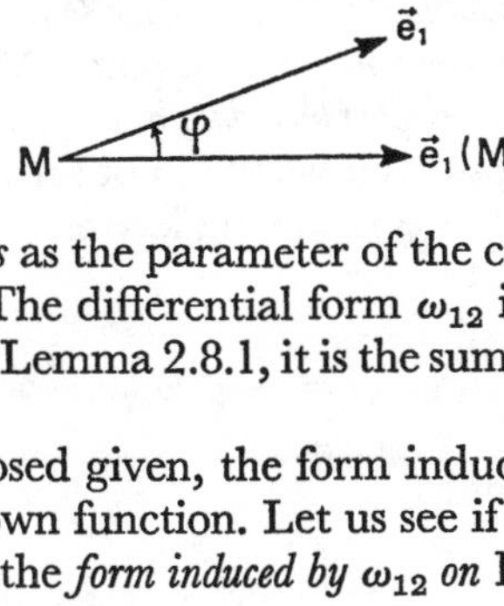

Let us use the arc length s as the parameter of the curve C. Then $\Gamma \subset \mathrm{R(S)} \approx \mathrm{S} \times \mathrm{SO}(2)$ is parametrized by s. The differential form ω_{12} induces on Γ a differential form which is easy to calculate: by Lemma 2.8.1, it is the sum of the forms induced by $\bar{\omega}_{12}$ and by $d\varphi$.

The curve C being supposed given, the form induced by $\bar{\omega}_{12}$ is known; it is of the type $a(s)\, ds$, where a is a known function. Let us see if it is possible to choose the functional form of $\varphi(s)$ such that the *form induced by* ω_{12} *on* Γ *is zero.* The condition is

$$d\varphi + a(s)\, ds = 0. \tag{2.9.1}$$

Thus it is necessary and sufficient that $\varphi(s)$ be a primitive of $-a(s)$. Thus the problem is possible, and the function φ is unique save for an arbitrary additive constant.

DEFINITION. If φ satisfies (2.9.1), we say that the field of vectors $\vec{e}_1$ at the points of C is a *parallel field* along C. By the above, a parallel field along a given curve C is defined by the (arbitrary) choice of a unit tangent vector at a point $\mathrm{M}_0 \in \mathrm{C}$; we then say that at an arbitrary point $\mathrm{M} \in \mathrm{C}$ the vector of the field is obtained *by parallel transport*, along C, of the vector given at the point M_0.

Parallel transport furnishes a simple interpretation of the *geodesic curvature* of an oriented curve C. Suppose that, at each point $\mathrm{M} \in \mathrm{C}$, θ is the angle between the unit tangent vector to C and the vector $\vec{e}_1(\mathrm{M})$ of the field of frames. This tangent vector is that of the Darboux frame of the curve C; the Darboux frames form a curve in R(S) ("canonical lifting" of C in R(S)), and the form induced by ω_{12} on this curve is equal to

$$\bar{\omega}_{12} + d\theta = a(s)\,ds + d\theta = d(\theta - \varphi),$$

φ defining as above a *parallel field* along C. Now we know that the form induced by ω_{12} on the canonical lifting of C is equal to ds/ρ_g, where $1/\rho_g$ is the geodesic curvature of C. Hence,

$$\boxed{\frac{1}{\rho_g} = \frac{d(\theta - \varphi)}{ds}}. \tag{2.9.2}$$

The geodesic curvature of C is equal to the derivative, with respect to the arc length s, of the angle $\theta - \varphi$ that the oriented tangent of C makes with the vector of a parallel field along C.

In particular, the geodesics are the curves C such that the field of unit tangent vectors at the points of C is a parallel field.

2.10 *Relation between total curvature and parallel transport*

As before, we suppose that there exists a field of frames on the surface S.

Let C be a *loop* (therefore the extremity of C coincides with its origin M_0). Choose a unit tangent vector to S at the point M_0, and effect the parallel transport of this vector along C. *It is not at all certain that we shall return to the point* M_0 *with the vector equal to its initial value*, even if the loop is homotopic to a point. More precisely: let γ be a loop, piecewise of class C^1, which is the *oriented boundary of a compact* δ. (N.B.: this implies that δ is homeomorphic to a compact disc, but we shall not prove this here. We see, therefore, that γ is homotopic to a point.) If φ is the angle which defines the parallel transport along γ, it follows from § 2.9 that

$$\int_\gamma d\varphi = -\int_\gamma \bar{\omega}_{12}.$$

By Stokes's theorem, this is equal to

$$-\iint_\delta d\bar{\omega}_{12},$$

therefore, by (2.8.2), to

$$\iint_\delta \frac{d\sigma}{R_1 R_2}.$$

Hence we have the fundamental relation

$$\boxed{\int_\gamma d\varphi = \iint_\delta \frac{d\sigma}{R_1 R_2}}. \tag{2.10.1}$$

The left-hand side is the *total angle turned through by the vector of the parallel field along the loop* γ (the angle φ is measured, at each point, with respect to the frame of the field of frames). We can therefore say that parallel transport along the loop γ *produces a rotation through an angle equal to the double integral, over* $\mathcal{S}$, *of the total curvature* $1/R_1R_2$ *with respect to the element of area* $d\sigma$.

Let us use the notation of § 2.9: θ denotes the angle between the unit tangent vector to γ at the point M and the vector $\vec{e}_1(M)$ of the field of frames. By (2.9.2) we have:

$$d\varphi = d\theta - \frac{ds}{\rho_g},$$

so that (2.10.1) becomes

$$\int_\gamma \frac{ds}{\rho_g} = \int_\gamma d\theta - \iint_{\mathcal{S}} \frac{d\sigma}{R_1R_2}. \tag{2.10.2}$$

To write this, we have implicitly assumed that γ is a curve of class C^2. If γ is composed of a finite number of arcs γ_i of class C^2, separated by angular points P_i where the angle of the unit tangent vector to γ has a discontinuity θ_i, the relation (2.10.2) must be set in the form:

$$\sum_i \int_{\gamma_i} \frac{ds}{\rho_g} = \sum_i \int_{\gamma_i} d\theta - \iint_{\mathcal{S}} \frac{d\sigma}{R_1R_2}. \tag{2.10.3}$$

We shall state without proof the following lemma, which introduces the topological properties of the plane:

Lemma 2.10.1. With the preceding notation, we have

$$\sum_i \int_{\gamma_i} d\theta + \sum_i \theta_i = 2\pi.$$

Intuitively, the left-hand side is the *total angle turned through by the unit tangent vector to* γ, reckoned at each point with respect to the vector $\vec{e}_1(M)$ of the field of frames. *A priori*, the left-hand side must be an integral multiple of 2π, since at each instant the angle θ is defined modulo 2π. The lemma affirms that it is exactly 2π.

The reader may verify this in the case where the surface S is a plane, $\mathcal{S}$ a disc, and γ is the bounding oriented circle of this disc: the tangent to the circle turns through 2π when the circle is described in the direct sense. The proof of the general case, roughly speaking, is deduced from this by suitable deformations.

Lemma 2.10.1 and (2.10.3) give:

THEOREM 2.10.2 (Gauss–Bonnet). *Under the preceding hypotheses:*

$$\boxed{\sum_i \int_{\gamma_i} \frac{ds}{\rho_g} + \sum_i \theta_i = 2\pi - \iint_{\mathcal{S}} \frac{d\sigma}{R_1R_2}}. \tag{2.10.4}$$

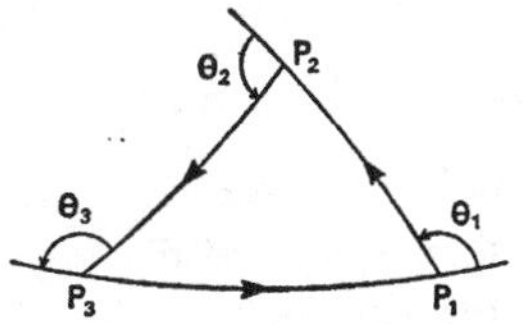

Particular case: let us suppose that γ is a *geodesic triangle*, i.e., that γ consists of three geodesic arcs $\gamma_1, \gamma_2, \gamma_3$. Let $\alpha_1, \alpha_2, \alpha_3$ denote the angles of this triangle (lying between 0 and 2π). Then evidently

$$\theta_i = \pi - \alpha_i,$$

whereas

$$\int_{\gamma_i} \frac{ds}{\rho_g} = 0,$$

since the geodesic curvature of γ_i is zero. Then (2.10.4) gives

$$3\pi - (\alpha_1 + \alpha_2 + \alpha_3) = 2\pi - \iint_\delta \frac{d\sigma}{R_1R_2},$$

i.e.,

$$\alpha_1 + \alpha_2 + \alpha_3 = \pi + \iint_\delta \frac{d\sigma}{R_1R_2}. \tag{2.10.5}$$

Hence:

COROLLARY 2.10.5. *The sum of the angles of a geodesic triangle, diminished by* π, *is equal to the integral of the total curvature taken over the interior of the triangle.*

Example: spherical triangles. Let S_2 be the unit sphere of $\mathbf{R}^3$. The total curvature is equal to 1 at every point of S_2 (every point is umbilical, with curvature equal to 1). A "spherical triangle" is, by definition, a geodesic triangle on S_2: its sides are arcs of *great circles.* There always exists a "field of frames" on a spherical triangle (indeed, there exists a point $P \in S_2$ which does not belong to the triangle; a stereographic projection with pole P transforms a plane into $S_2 - \{P\}$, and a field of vectors $\neq 0$ in the plane into a field of vectors $\neq 0$ onto $S_2 - \{P\}$). We can thus apply Corollary 2.10.3: *the sum of the angles of a spherical triangle, diminished by* π, *is equal to the area of the triangle.* This result can also be proved by elementary methods of geometry (see Hadamard's elementary *Treatise on Geometry*).

For example, consider the spherical triangle formed by the intersection of S_2 with the trihedral tri-rectangle having its vertex at the centre of the sphere. Each angle of the triangle is a right angle; thus its area is equal to $\pi/2$. This is in agreement with the fact that the sphere S_2 is the union of *eight* such triangles.

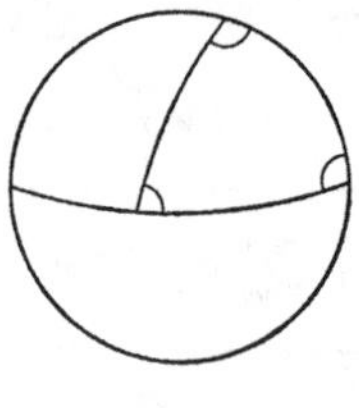

2.11 *Calculation of the total curvature of a surface by means of the first fundamental form*

Let us suppose that we have a field of frames on S, and proceed in terms of the notation of § 2.8. Since the differential forms $\bar{\omega}_1$ and $\bar{\omega}_2$ are linearly independent at each point of S, the form $\bar{\omega}_{12}$ can be written uniquely as

$$\bar{\omega}_{12} = \lambda_1\bar{\omega}_1 + \lambda_2\bar{\omega}_2. \tag{2.11.1}$$

We shall calculate the functions λ_1 and λ_2 by means of (2.8.1). Indeed (2.11.1) and (2.8.1) give

$$d\bar{\omega}_1 = \lambda_1\bar{\omega}_1 \wedge \bar{\omega}_2, \qquad d\omega_2 = \lambda_2\bar{\omega}_1 \wedge \bar{\omega}_2. \tag{2.11.2}$$

If we know $\bar{\omega}_1$ and $\bar{\omega}_2$, we can calculate $d\bar{\omega}_1$ and $d\bar{\omega}_2$, and then λ_1 and λ_2 (functions on S) are given by (2.11.2). Next, knowing $\bar{\omega}_{12}$, we calculate $d\bar{\omega}_{12}$, and this gives the total curvature in virtue of

$$d\bar{\omega}_{12} = -\frac{1}{R_1R_2}\bar{\omega}_1 \wedge \bar{\omega}_2 \qquad \text{(cf. (2.8.2))}. \tag{2.11.3}$$

Thus it reduces to the calculation of the differential forms $\bar{\omega}_1$ and $\bar{\omega}_2$. These are functions $\bar{\omega}_i(M, \vec{\xi})$, $\vec{\xi}$ being tangential to S at the point M: for given M, $\bar{\omega}_1(M, \vec{\xi})$ and $\bar{\omega}_2(M, \vec{\xi})$ are the coordinates of the vector $\vec{\xi}$ with respect to the base $(\vec{e}_1(M), \vec{e}_2(M))$.

Let us carry out the complete computations when we have a representation in terms of parameters (u, v) such that

$$ds^2 = (A\,du)^2 + (B\,dv)^2, \tag{2.11.4}$$

where A and B are functions of (u, v) which are > 0.

On S take the field of vectors which corresponds to the field of constant vectors $(1, 0)$ in the (u, v)-plane. Then

$$\bar{\omega}_1 = A\,du, \qquad \bar{\omega}_2 = B\,dv.$$

We have

$$d\bar{\omega}_1 = -\frac{\partial A}{\partial v}du \wedge dv, \qquad d\omega_2 = \frac{\partial B}{\partial u}du \wedge dv, \qquad \bar{\omega}_1 \wedge \bar{\omega}_2 = AB\,du \wedge dv,$$

hence, substituting in (2.11.2):

$$\lambda_1 = -\frac{1}{AB}\frac{\partial A}{\partial v}, \qquad \lambda_2 = \frac{1}{AB}du \wedge dv.$$

Then (2.11.1) gives

$$\bar{\omega}_{12} = -\frac{1}{B}\frac{\partial A}{\partial v}du + \frac{1}{A}\frac{\partial B}{\partial u}dv,$$

hence

$$d\bar{\omega}_{12} = \left[\frac{\partial}{\partial v}\left(\frac{1}{B}\frac{\partial A}{\partial v}\right) + \frac{\partial}{\partial u}\left(\frac{1}{A}\frac{\partial B}{\partial u}\right)\right]du \wedge dv,$$

and (2.11.3) finally gives

$$\frac{1}{R_1R_2} = -\frac{1}{AB}\left[\frac{\partial}{\partial v}\left(\frac{1}{B}\frac{\partial A}{\partial v}\right) + \frac{\partial}{\partial u}\left(\frac{1}{A}\frac{\partial B}{\partial u}\right)\right]. \tag{2.11.5}$$

This is the formula for the total curvature when ds^2 is in the form (2.11.4).

Exercises

Exercises on the moving frame method

The notation employed will be the same as that of the last chapter.

A surface (S) will denote a connected differential variety of 2 dimensions, and of class $C^k (k \geqslant 2)$ in $\mathbf{R}^3$.

1. Show that the normal curvature in the direction of the vector $\vec{u}$ lying in the tangent plane at the point M of a surface (S) is given by

$$\rho_n = \frac{1}{R_n} = \frac{a\omega_1^2 + 2b\omega_1\omega_2 + c\omega_2^2}{\omega_1^2 + \omega_2^2},$$

the forms ω_1 and ω_2 being calculated on $\vec{u}$.

Show that the principal directions at M are those for which $1/R_n$ is extremal. Deduce that for a principal direction $\vec{u}$ at M,

$$\frac{1}{R} = \frac{a\omega_1 + b\omega_2}{\omega_1} = \frac{b\omega_1 + c\omega_2}{\omega_2},$$

$1/R$ being the principal curvature relative to $\vec{u}$, the forms ω_1 and ω_2 being calculated on $\vec{u}$.

Deduce further that the differential equation of the lines of curvature on (S) is

$$b(\omega_1^2 - \omega_2^2) + (c - a)\omega_1\omega_2 = 0.$$

2. Obtain the form of the equations of the moving frame when $\vec{e}_1$ and $\vec{e}_2$ are unit vectors in the principal directions at the point M. *Such a frame will be used systematically in the following*; a, b, c will therefore denote functions on the corresponding family of frames, or, what amounts to the same thing, on the surface S.

3. A point M of (S) such that $a = b = c = 0$ is called a flat point. Show that if every point of a closed surface (S) is flat, then (S) is a plane.

4. Using the frame of Ex. 2, study the surfaces (S) whose total curvature $K = ac - b^2$ is zero.

We can show that ω_{13} or ω_{23} is zero. Consider the case $\omega_{23} \neq 0$, $\omega_{13} = 0$, show that locally there exists on (S) a function u such that $\omega_{23} = du$, and study the displacement of a moving frame along the lines $u =$ const. Deduce that (S) is a developable surface. In the particular case $\omega_{12} = 0$, show that (S) is a cylinder.

5. A point M of (S) such that $a = c = \rho \neq 0$ and $b = 0$ is called an umbilical point. Let (S) be a closed surface of which every point is umbilical. Show that $d\rho \wedge \omega_1 = d\rho \wedge \omega_2 = 0$ and deduce that (S) is a sphere of radius $1/|\rho|$.

6. Show that a surface (S) whose mean curvature $H = a + c$ is equal to 1 and whose total curvature $K = ac - b^2$ is zero is a right circular cylinder of radius 1.

By Ex. 2 and 4, we have the case where $\omega_{13} = 0$.

Show that locally there exist on (S) independent functions u and v such that

$$\begin{cases} d\vec{M} = du\vec{e}_1 + dv\vec{e}_2; & d\vec{e}_2 = dv\vec{e}_3 \\ d\vec{e}_1 = 0 & d\vec{e}_3 = -dv\vec{e}_2 \end{cases}$$

and comment.

7. Show that on a compact surface (S) there exists a point M where the total curvature is strictly positive. To do this, if 0 is a point of $\mathbf{R}^3$ and M a point of (S) where 0M admits of a maximum, it may be shown that the second fundamental form of (S) is negative definite at the point M.

If the mean curvature $H = a + c$ is zero at every point, (S) is called a minimal surface. Are there any compact minimal surfaces?

8. Considering the frame of Ex. 2, show that if $da = dc = 0$ at the point M, then at M either $a = c$, or $\omega_{12} = 0$. Deduce that on a surface (S) which has a constant strictly positive total curvature K, the principal curvatures cannot have a relative maximum or minimum at a point which is not umbilical.

9. Using Ex. 8 and 5, show that a compact surface (S) with constant strictly positive total curvature is a sphere.

10. If a closed surface (S) has no umbilical points, and if $H = a + c$ and $K = ac - b^2$ are constant, then (S) is a right circular cylinder. Are there any minimal surfaces which have constant, strictly negative total curvature? Use Ex. 8 and 6.

11. *Spherical transformation.* With every point M of (S) we identify the point $\mu \in \Sigma$, the sphere centre 0 and radius 1, such that $\vec{0\mu} = \vec{e}_3$. The mapping $f\colon S \to \Sigma$ such that $f(M) = \mu$ thus defined is called the spherical transformation of (S) into (Σ).

(a) The tangent planes $T_M(S)$ of (S) at M, and $T_\mu(\Sigma)$ of (Σ) at μ being parallel, the first fundamental form of (Σ) at μ naturally induces a quadratic form on $T_M(S)$. Calculate this quadratic form as a function of a, b, c, ω_1, ω_2. The quadratic form thus obtained will be called the 3rd fundamental quadratic form of (S) at M.

(b) Similarly the derived mapping f' of f at M, induces a linear mapping L of the tangent plane $T_M(S)$ into itself. L is called the Weingarten mapping of (S) at M. Using the formulae for $d\vec{M}$ and $d\vec{e}_3$, calculate the matrix of L with respect to the base $(\vec{e}_1, \vec{e}_2)$ of $T_M(S)$ as a function of a, b, c and deduce that L is a symmetric operator for the first fundamental form of (S) at M.

(c) Calculate the eigenvalues, and the eigenvectors, the trace and the determinant of L, and interpret geometrically the results obtained. Give a condition which must be satisfied if f is to be a local diffeomorphism of (S) onto (Σ).

(d) Denoting by I, II, III the first, second and third fundamental forms of (S) at M respectively, show that $\forall \vec{X}, \vec{Y} \in T_M(S)$ we have

$$\mathrm{II}(\vec{X}, \vec{Y}) = -\mathrm{I}(L\cdot\vec{X}, \vec{Y}) = -\mathrm{I}(\vec{X}, L\cdot\vec{Y})$$

$$\mathrm{III}(\vec{X}, \vec{Y}) = \mathrm{I}(L^2\cdot\vec{X}, \vec{Y}) = \mathrm{I}(L\cdot\vec{X}, L\cdot\vec{Y}) = \mathrm{I}(\vec{X}, L^2\cdot\vec{Y})$$

(e) $H = a + c$ and $K = ac - b$ being respectively the mean curvature and the total curvature of (S) at M, show that:

$$L^2 + HL + K\cdot Id = 0 \qquad (Id\text{, the identity mapping of } T_M(S)).$$

Deduce that $\mathrm{III} - H\cdot\mathrm{II} + K\cdot\mathrm{I} = 0$.

12. (Depends on 11.) The spherical mapping $f\colon S \to \Sigma$ is strictly conformal if $\forall \vec{X}, \vec{Y} \in T_M(S)$, there exists a function $u(M)$ which is strictly positive on (S) such that

$$\mathrm{I}(L\cdot\vec{X}, L\cdot\vec{Y}) = \mathrm{III} \qquad (\vec{X}, \vec{Y}) = u(M)\mathrm{I}(\vec{X}, \vec{Y}).$$

Show that if (S) is compact, (S) is a sphere, and if (S) is not compact, (S) is a minimal surface with strictly negative total curvature equal to $-u(M)$.

(The corresponding relations between a, b, c are interpreted by using the results of Ex. 5 and 7.)

13. (Depends on 12.) Show that if I = II or I = III, if (S) is closed, (S) is a sphere of radius 1, and conversely. (If II = III, show that if (S) is closed, (S) is a sphere of radius 1, a plane or a right circular cylinder of radius 1.)

14. *Parallel surfaces.* Consider a surface (S); let (S_r) be the locus of the point $\varphi(M) = M_r = M + r\vec{e}_3$, where $M \in (S)$ and r is constant.
(a) Consider the frame $(M_r, \vec{e}_1', \vec{e}_2', \vec{e}_3')$ obtained from the frame $(M, \vec{e}_1, \vec{e}_2, \vec{e}_3)$ by the translation $r\vec{e}_3$. Calculate the forms $\bar{\omega}_i$ and $\bar{\omega}_{ij}$ of this new frame on (S_r) as functions of the forms ω_i and ω_{ij} of the frame $(M, \vec{e}_1, \vec{e}_2, \vec{e}_3)$.
(b) Show that the principal directions, the principal curvatures, the mean curvature, the total curvature and the three fundamental forms of (S_r) at M_r can be obtained as functions of the corresponding elements of (S) at M provided a certain relation between H, K and r is satisfied.
(c) The tangent planes $T_M(S)$ and $T_{M_r}(S_r)$ being parallel, the derived mapping φ' of φ naturally induces a linear mapping F of $T_M(S)$ into itself; obtain the matrix of this mapping in terms of the base $\vec{e}_1, \vec{e}_2$ as a function of a, b, c, and r. Obtain as a function of F the condition found in (b), and show that in this case φ is a local diffeomorphism of (S) onto (S_r). The Weingarten mapping (Ex. 11) of S_r at M_r also induces a linear mapping of $T_M(S)$. If L is the Weingarten mapping of (S) at M, show that $L_r \cdot F = F \cdot L_r = L$ and rederive the results of (c).

15. Show that if the mapping $\varphi: S \to S_r$ (Ex. 14) is strictly conformal and if further (S) is closed, then (S) is a sphere or a plane or a surface without umbilical points with constant mean curvature $H = 2/r$.

16. A triple-orthogonal system defined in an open U of $\mathbf{R}^3$ is a family of surfaces such that through each point M of U pass exactly 3 members of the family whose normals at M are mutually orthogonal. Show that the surfaces of a triple orthogonal system intersect along their lines of curvature.

To do this, consider the displacement of an orthonormal moving frame $(M, \vec{e}_1, \vec{e}_2, \vec{e}_3)$, $M \in U$ and $\vec{e}_1, \vec{e}_2, \vec{e}_3$ are respectively normals to the 3 surfaces S_1, S_2, S_3 passing through M.

As in § 1.1, put

$$dM = \sum_{i=1}^{3} \omega_i \vec{e}_i \quad \text{and} \quad d\vec{e}_i = \sum_{j=1}^{3} \omega_{ij}\vec{e}_j \qquad (i = 1, 2, 3).$$

Show that we can write

$$\begin{aligned} \omega_{23} &= a_{11}\omega_1 + a_{12}\omega_2 + a_{13}\omega_3 \\ \omega_{31} &= a_{21}\omega_1 + a_{22}\omega_2 + a_{23}\omega_3 \\ \omega_{12} &= a_{31}\omega_1 + a_{32}\omega_2 + a_{33}\omega_3 \end{aligned}$$

next show that, if M is constrained to move on (S_3), $a_{11} = -a_{22}$. Deduce that $a_{11} = a_{22} = a_{33} = 0$ and use Ex. 2 to show that $(\vec{e}_1, \vec{e}_2, \vec{e}_3)$ are the principal directions to the surfaces (S_1), (S_2), (S_3) at r.

17. Show that a family of concentric spheres is naturally a part of an infinity of triple-orthogonal systems, and deduce that a conformal diffeomorphism of $\mathbf{R}^3$ transforms the spheres into spheres.

Exercises using integration

18. Using equations (1.4.4) show that if the torsion and the curvature of a curve C of $\mathbf{R}^3$ are proportional ($\rho/\tau = k$) the curve C is a helix for a prescribed direction $\vec{u}$ (we say that a curve is a helix for a prescribed direction $\vec{u}$, which is fixed, if its tangent makes a constant angle with $\vec{u}$).

Particular case. Suppose $\rho = \tau = (1 + s)\sqrt{2}$. Show that C is defined to within a translation, and find its equation, when the initial conditions, at $s = 0$, are

$$\vec{e}_1 = \left(\frac{1}{2}, \frac{1}{2}, \frac{1}{\sqrt{2}}\right), \qquad \vec{e}_2 = \left(-\frac{1}{\sqrt{2}}, \frac{1}{\sqrt{2}}, 0\right), \qquad \vec{e}_3 = \left(-\frac{1}{2}, -\frac{1}{2}, \frac{1}{\sqrt{2}}\right).$$

19. Using the frame of Ex. 2, show that if $\omega_1 = du$, i.e., that the curves $v =$ const. are, at the same time, lines of curvature and geodesics, cf. Ex. 14 of Chap. 2, then the curves are plane (show that ω_{12} is proportional to ω_2).

In particular consider the surfaces of revolution, and show that, in this case, a and c are functions of u alone.

Conversely, suppose that, for a surface (S), $\omega_1 = du$, $a = \varphi(u)$, $c = \psi(u)$. Show that the parameter v may be chosen in such a manner that $\omega_2 = \mathrm{H}(u)\,dv$, where H depends only on u.

Show that

$$\omega_{12} = \frac{\mathrm{H}'(u)}{\mathrm{H}(u)}\,\omega_2, \quad \text{and that} \quad \mathrm{H}'^2 + \mathrm{C}^2\mathrm{H}^2 = \text{const.}$$

Integrate the system $d\vec{e}_i = \sum_j \omega_{ij}\vec{e}_j$, then $d\vec{\mathrm{M}} = \sum_i \omega_i\vec{e}_i$, and show that (S) is a surface of revolution.

20. Using the method of the preceding exercise, find a surface (S) such that $\omega_1 = du$, $a = \varphi(u)$, $c = \psi(u)$, with $\varphi(u) + \psi(u) = 0$.

Show that the meridian of this surface of revolution is a catenary.

Index

Bibliography

In *Differential Forms*, as well as in the preceding *Differential Calculus*, the author was inspired by the excellent treatise of Jean Dieudonné, *Foundations of Modern Analysis*, Academic Press, New York, 1960, and especially by Chapters V, VIII and X.

He has also made profitable use of *Introduction to Differentiable Manifolds*, by Serge Lang, Interscience Publishers, New York, 1962, and of the *Notes on Differential Calculus* of Professor Lelong-Ferrand, published by the Centre de documentation universitaire (France).

A CATALOG OF SELECTED

DOVER BOOKS

IN SCIENCE AND MATHEMATICS

Mathematics

FUNCTIONAL ANALYSIS (Second Corrected Edition), George Bachman and Lawrence Narici. Excellent treatment of subject geared toward students with background in linear algebra, advanced calculus, physics and engineering. Text covers introduction to inner-product spaces, normed, metric spaces, and topological spaces; complete orthonormal sets, the Hahn-Banach Theorem and its consequences, and many other related subjects. 1966 ed. 544pp. $6\frac{1}{8}$ x $9\frac{1}{4}$. 0-486-40251-7

ASYMPTOTIC EXPANSIONS OF INTEGRALS, Norman Bleistein & Richard A. Handelsman. Best introduction to important field with applications in a variety of scientific disciplines. New preface. Problems. Diagrams. Tables. Bibliography. Index. 448pp. $5\frac{3}{8}$ x $8\frac{1}{2}$. 0-486-65082-0

VECTOR AND TENSOR ANALYSIS WITH APPLICATIONS, A. I. Borisenko and I. E. Tarapov. Concise introduction. Worked-out problems, solutions, exercises. 257pp. $5\frac{3}{8}$ x $8\frac{1}{4}$. 0-486-63833-2

AN INTRODUCTION TO ORDINARY DIFFERENTIAL EQUATIONS, Earl A. Coddington. A thorough and systematic first course in elementary differential equations for undergraduates in mathematics and science, with many exercises and problems (with answers). Index. 304pp. $5\frac{3}{8}$ x $8\frac{1}{2}$. 0-486-65942-9

FOURIER SERIES AND ORTHOGONAL FUNCTIONS, Harry F. Davis. An incisive text combining theory and practical example to introduce Fourier series, orthogonal functions and applications of the Fourier method to boundary-value problems. 570 exercises. Answers and notes. 416pp. $5\frac{3}{8}$ x $8\frac{1}{2}$. 0-486-65973-9

COMPUTABILITY AND UNSOLVABILITY, Martin Davis. Classic graduate-level introduction to theory of computability, usually referred to as theory of recurrent functions. New preface and appendix. 288pp. $5\frac{3}{8}$ x $8\frac{1}{2}$. 0-486-61471-9

ASYMPTOTIC METHODS IN ANALYSIS, N. G. de Bruijn. An inexpensive, comprehensive guide to asymptotic methods–the pioneering work that teaches by explaining worked examples in detail. Index. 224pp. $5\frac{3}{8}$ x $8\frac{1}{2}$ 0-486-64221-6

APPLIED COMPLEX VARIABLES, John W. Dettman. Step-by-step coverage of fundamentals of analytic function theory–plus lucid exposition of five important applications: Potential Theory; Ordinary Differential Equations; Fourier Transforms; Laplace Transforms; Asymptotic Expansions. 66 figures. Exercises at chapter ends. 512pp. $5\frac{3}{8}$ x $8\frac{1}{2}$. 0-486-64670-X

INTRODUCTION TO LINEAR ALGEBRA AND DIFFERENTIAL EQUATIONS, John W. Dettman. Excellent text covers complex numbers, determinants, orthonormal bases, Laplace transforms, much more. Exercises with solutions. Undergraduate level. 416pp. $5\frac{3}{8}$ x $8\frac{1}{2}$. 0-486-65191-6

RIEMANN'S ZETA FUNCTION, H. M. Edwards. Superb, high-level study of landmark 1859 publication entitled "On the Number of Primes Less Than a Given Magnitude" traces developments in mathematical theory that it inspired. xiv+315pp. $5\frac{3}{8}$ x $8\frac{1}{2}$. 0-486-41740-9

CATALOG OF DOVER BOOKS

INTRODUCTORY REAL ANALYSIS, A.N. Kolmogorov, S. V. Fomin. Translated by Richard A. Silverman. Self-contained, evenly paced introduction to real and functional analysis. Some 350 problems. 403pp. $5\frac{3}{8}$ x $8\frac{1}{2}$. 0-486-61226-0

APPLIED ANALYSIS, Cornelius Lanczos. Classic work on analysis and design of finite processes for approximating solution of analytical problems. Algebraic equations, matrices, harmonic analysis, quadrature methods, much more. 559pp. $5\frac{3}{8}$ x $8\frac{1}{2}$. 0-486-65656-X

AN INTRODUCTION TO ALGEBRAIC STRUCTURES, Joseph Landin. Superb self-contained text covers "abstract algebra": sets and numbers, theory of groups, theory of rings, much more. Numerous well-chosen examples, exercises. 247pp. $5\frac{3}{8}$ x $8\frac{1}{2}$. 0-486-65940-2

QUALITATIVE THEORY OF DIFFERENTIAL EQUATIONS, V. V. Nemytskii and V.V. Stepanov. Classic graduate-level text by two prominent Soviet mathematicians covers classical differential equations as well as topological dynamics and ergodic theory. Bibliographies. 523pp. $5\frac{3}{8}$ x $8\frac{1}{2}$. 0-486-65954-2

THEORY OF MATRICES, Sam Perlis. Outstanding text covering rank, nonsingularity and inverses in connection with the development of canonical matrices under the relation of equivalence, and without the intervention of determinants. Includes exercises. 237pp. $5\frac{3}{8}$ x $8\frac{1}{2}$. 0-486-66810-X

INTRODUCTION TO ANALYSIS, Maxwell Rosenlicht. Unusually clear, accessible coverage of set theory, real number system, metric spaces, continuous functions, Riemann integration, multiple integrals, more. Wide range of problems. Undergraduate level. Bibliography. 254pp. $5\frac{3}{8}$ x $8\frac{1}{2}$. 0-486-65038-3

MODERN NONLINEAR EQUATIONS, Thomas L. Saaty. Emphasizes practical solution of problems; covers seven types of equations. ". . . a welcome contribution to the existing literature...."–*Math Reviews*. 490pp. $5\frac{3}{8}$ x $8\frac{1}{2}$. 0-486-64232-1

MATRICES AND LINEAR ALGEBRA, Hans Schneider and George Phillip Barker. Basic textbook covers theory of matrices and its applications to systems of linear equations and related topics such as determinants, eigenvalues and differential equations. Numerous exercises. 432pp. $5\frac{3}{8}$ x $8\frac{1}{2}$. 0-486-66014-1

LINEAR ALGEBRA, Georgi E. Shilov. Determinants, linear spaces, matrix algebras, similar topics. For advanced undergraduates, graduates. Silverman translation. 387pp. $5\frac{3}{8}$ x $8\frac{1}{2}$. 0-486-63518-X

ELEMENTS OF REAL ANALYSIS, David A. Sprecher. Classic text covers fundamental concepts, real number system, point sets, functions of a real variable, Fourier series, much more. Over 500 exercises. 352pp. $5\frac{3}{8}$ x $8\frac{1}{2}$. 0-486-65385-4

SET THEORY AND LOGIC, Robert R. Stoll. Lucid introduction to unified theory of mathematical concepts. Set theory and logic seen as tools for conceptual understanding of real number system. 496pp. $5\frac{3}{8}$ x $8\frac{1}{4}$. 0-486-63829-4

Physics

OPTICAL RESONANCE AND TWO-LEVEL ATOMS, L. Allen and J. H. Eberly. Clear, comprehensive introduction to basic principles behind all quantum optical resonance phenomena. 53 illustrations. Preface. Index. 256pp. 5⅜ x 8½. 0-486-65533-4

QUANTUM THEORY, David Bohm. This advanced undergraduate-level text presents the quantum theory in terms of qualitative and imaginative concepts, followed by specific applications worked out in mathematical detail. Preface. Index. 655pp. 5⅜ x 8½. 0-486-65969-0

ATOMIC PHYSICS (8th EDITION), Max Born. Nobel laureate's lucid treatment of kinetic theory of gases, elementary particles, nuclear atom, wave-corpuscles, atomic structure and spectral lines, much more. Over 40 appendices, bibliography. 495pp. 5⅜ x 8½. 0-486-65984-4

A SOPHISTICATE'S PRIMER OF RELATIVITY, P. W. Bridgman. Geared toward readers already acquainted with special relativity, this book transcends the view of theory as a working tool to answer natural questions: What is a frame of reference? What is a "law of nature"? What is the role of the "observer"? Extensive treatment, written in terms accessible to those without a scientific background. 1983 ed. xlviii+172pp. 5⅜ x 8½. 0-486-42549-5

AN INTRODUCTION TO HAMILTONIAN OPTICS, H. A. Buchdahl. Detailed account of the Hamiltonian treatment of aberration theory in geometrical optics. Many classes of optical systems defined in terms of the symmetries they possess. Problems with detailed solutions. 1970 edition. xv + 360pp. 5⅜ x 8½. 0-486-67597-1

PRIMER OF QUANTUM MECHANICS, Marvin Chester. Introductory text examines the classical quantum bead on a track: its state and representations; operator eigenvalues; harmonic oscillator and bound bead in a symmetric force field; and bead in a spherical shell. Other topics include spin, matrices, and the structure of quantum mechanics; the simplest atom; indistinguishable particles; and stationary-state perturbation theory. 1992 ed. xiv+314pp. 6⅛ x 9¼. 0-486-42878-8

LECTURES ON QUANTUM MECHANICS, Paul A. M. Dirac. Four concise, brilliant lectures on mathematical methods in quantum mechanics from Nobel Prize-winning quantum pioneer build on idea of visualizing quantum theory through the use of classical mechanics. 96pp. 5⅜ x 8½. 0-486-41713-1

THIRTY YEARS THAT SHOOK PHYSICS: THE STORY OF QUANTUM THEORY, George Gamow. Lucid, accessible introduction to influential theory of energy and matter. Careful explanations of Dirac's anti-particles, Bohr's model of the atom, much more. 12 plates. Numerous drawings. 240pp. 5⅜ x 8½. 0-486-24895-X

ELECTRONIC STRUCTURE AND THE PROPERTIES OF SOLIDS: THE PHYSICS OF THE CHEMICAL BOND, Walter A. Harrison. Innovative text offers basic understanding of the electronic structure of covalent and ionic solids, simple metals, transition metals and their compounds. Problems. 1980 edition. 582pp. 6⅛ x 9¼. 0-486-66021-4

A TREATISE ON ELECTRICITY AND MAGNETISM, James Clerk Maxwell. Important foundation work of modern physics. Brings to final form Maxwell's theory of electromagnetism and rigorously derives his general equations of field theory. 1,084pp. $5\frac{3}{8}$ x $8\frac{1}{2}$. Two-vol. set. Vol. I: 0-486-60636-8 Vol. II: 0-486-60637-6

QUANTUM MECHANICS: PRINCIPLES AND FORMALISM, Roy McWeeny. Graduate student-oriented volume develops subject as fundamental discipline, opening with review of origins of Schrödinger's equations and vector spaces. Focusing on main principles of quantum mechanics and their immediate consequences, it concludes with final generalizations covering alternative "languages" or representations. 1972 ed. 15 figures. xi+155pp. $5\frac{3}{8}$ x $8\frac{1}{2}$. 0-486-42829-X

INTRODUCTION TO QUANTUM MECHANICS With Applications to Chemistry, Linus Pauling & E. Bright Wilson, Jr. Classic undergraduate text by Nobel Prize winner applies quantum mechanics to chemical and physical problems. Numerous tables and figures enhance the text. Chapter bibliographies. Appendices. Index. 468pp. $5\frac{3}{8}$ x $8\frac{1}{2}$. 0-486-64871-0

METHODS OF THERMODYNAMICS, Howard Reiss. Outstanding text focuses on physical technique of thermodynamics, typical problem areas of understanding, and significance and use of thermodynamic potential. 1965 edition. 238pp. $5\frac{3}{8}$ x $8\frac{1}{2}$. 0-486-69445-3

THE ELECTROMAGNETIC FIELD, Albert Shadowitz. Comprehensive undergraduate text covers basics of electric and magnetic fields, builds up to electromagnetic theory. Also related topics, including relativity. Over 900 problems. 768pp. $5\frac{5}{8}$ x $8\frac{1}{4}$. 0-486-65660-8

GREAT EXPERIMENTS IN PHYSICS: FIRSTHAND ACCOUNTS FROM GALILEO TO EINSTEIN, Morris H. Shamos (ed.). 25 crucial discoveries: Newton's laws of motion, Chadwick's study of the neutron, Hertz on electromagnetic waves, more. Original accounts clearly annotated. 370pp. $5\frac{3}{8}$ x $8\frac{1}{2}$. 0-486-25346-5

EINSTEIN'S LEGACY, Julian Schwinger. A Nobel Laureate relates fascinating story of Einstein and development of relativity theory in well-illustrated, nontechnical volume. Subjects include meaning of time, paradoxes of space travel, gravity and its effect on light, non-Euclidean geometry and curving of space-time, impact of radio astronomy and space-age discoveries, and more. 189 b/w illustrations. xiv+250pp. $8\frac{3}{8}$ x $9\frac{1}{4}$. 0-486-41974-6

STATISTICAL PHYSICS, Gregory H. Wannier. Classic text combines thermodynamics, statistical mechanics and kinetic theory in one unified presentation of thermal physics. Problems with solutions. Bibliography. 532pp. $5\frac{3}{8}$ x $8\frac{1}{2}$. 0-486-65401-X

Paperbound unless otherwise indicated. Available at your book dealer, online at **www.doverpublications.com**, or by writing to Dept. GI, Dover Publications, Inc., 31 East 2nd Street, Mineola, NY 11501. For current price information or for free catalogues (please indicate field of interest), write to Dover Publications or log on to **www.doverpublications.com** and see every Dover book in print. Dover publishes more than 500 books each year on science, elementary and advanced mathematics, biology, music, art, literary history, social sciences, and other areas.